CW00500409

FERRARI THE MAN, THE MACHINES

AN AUTOMOBILE QUARTERLY LIBRARY SERIES BOOK

Frederick Muller Limited
London

FERR

THE MAN, THE

Edited, and with an Introd

WITH CHAPTERS BY: Griffith Borgeson Stirling Mo

Phil Hill Karl Ludvigsen Peter C. Coltrin Stan Nowak

PUBLISHED BY FREDE

LO

ARI
MACHINES

...tion, by Stan Grayson

Denis Jenkinson Jan P. Norbye

...lbert R. Bochroch David Owen

...CK MULLER LIMITED

...DON

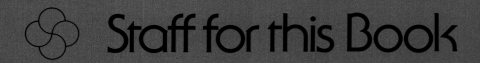

Staff for this Book

AUTOMOBILE QUARTERLY PUBLICATIONS

Publisher and President: L. Scott Bailey
Executive Editor: Beverly Rae Kimes
Book Editor: Stan Grayson
Art Director: Theodore R. F. Hall
Assistant Art Director: Edmond Fenech
Contributing Artists: Ken Rush, Yoshihiro Inomoto
Contributing Photographers: (black and white sections)
Jesse Alexander, Joe Bilbao, Griffith Borgeson,
Neill Bruce, Bernard Cahier, Geoffrey Goddard,
Karl Ludvigsen, Denise McCluggage, Corrado Millanta,
Julius Weitmann. (color sections) Giorgio Boschetti,
Neill Bruce, Henry Austin Clark Jr.,
Stan Grayson, Rick Lenz, Marc Madow.

This edition published in Great Britain by Frederick Muller Limited, Dataday House, London SW19 7JU

copyright © 1975 by Princeton Publishing Inc., Princeton New Jersey 08540

All rights reserved. No part of this publication may be reproduced, stored in a retrieval system, or transmitted, in any form or by any means, electronic, mechanical, photocopying, recording or otherwise, without the prior consent of Frederick Muller Limited.

ISBN 0-584-95027-6

Printed in Hong Kong by South China Printing Co.

Contents

Introduction

Ferrari. During almost three decades of carmaking, the name has acquired a magic beyond all others for those who love fast cars. Today Enzo Ferrari rests secure atop a pinnacle of his own making. He has persevered and has now achieved a status uniquely his own while behind him swirl the successes and failures of other car builders, engineers and financiers, the broken dreams of dead champions, all of which comprise the history of sporting automobiles and automobile racing. But who is Ferrari? What drives this single-minded and obviously brilliant man who has, in his lifetime, fostered tremendous victories in one of sport's most demanding and fabulous arenas and—at the same time—sparked tragedy, enormous creativity, a succession of beautiful objects, conspiracies of almost Shakespearean proportions? What is it, exactly, about the cars which Ferrari caused to be created that sets them so apart? There are no easy answers to these questions. A while back, we at AUTOMOBILE Quarterly set out to discover some of them, but almost at once the project grew beyond the scope of our magazine. There was so much ground to cover, so many pictures to include, so very much which needed to be presented that the result could simply not be set down within the confines of a single issue. So we decided on a book instead, and then we pondered its approach. An encyclopedia could be compiled about Ferrari's life and cars, the people he has known, those who have made their way from all over the world to his sanctum in Maranello. But an encyclopedia—with its data-laden, objectivity-bound and often stilted approach—isn't really equal to untangling a mystery or explaining a mystique. Nor did we believe that a single author, with a predetermined point of view, would be appropriate for the Ferrari book we envisioned. Rather, we concluded, an approach that would be subjective—in every meaning of the word—was the answer. And that is what is presented here. We have chosen a group of themes or topics, a variety of approaches to the kaleidoscopic creations of Ferrari's factory and the enigma of Ferrari himself. Each of the contributors to this book brings a

special expertise to his particular subject. The roster includes some of the world's best
automotive writers and a pair of its best racing drivers. At the heart of most of their efforts—the chapters that are
not primarily technical—is the concern to stress the human side of the Ferrari story, to name names, to tell something about the people
behind Ferrari automobiles. This has seldom been done before and for good reasons. Solid information
about Ferrari and his cars has always been at a premium and, as a matter of fact, it still is. Enzo Ferrari himself
has never been truly approachable and those few who do know him, or those who have worked
for him, have ever been reluctant to describe their impressions and
experiences for publication. Doubtless, it will be this behind-the-scenes story of Ferrari which will claim the
attention of journalists and automotive historians for years to come. Doubtless too, this is as it should be. For there have seldom been
cars which so forcefully communicate the fact that they are, after all, the
end-product of knowledgeable, living, breathing enthusiasts. Engineers with soul. This book, then, concerns itself with
an aura, with a technology and with the people behind that technology. It is a special subject
and we feel that the writers, artists, and photographers represented here have contributed something special to
the history of the high performance and racing automobile. The color photographs comprise—to our knowledge—the most comprehensive
selection of Ferrari car portraits ever published, as does the group of Formula One color drawings.
Taken together, we hope these essays, reminiscences and illustrations will present a fresh look at Ferrari and
answer some of the perplexing questions about him, both for longtime enthusiasts of the marque and for newcomers.
But enough introductory talk. What follows speaks for itself.

Stan Grayson
Princeton, New Jersey

CHAPTER 1

The Great Agitator

by Griffith Borgeson

Griffith Borgeson has been writing about automobiles—and the men who build and drive them—since 1949. Besides innumerable magazine articles, he authored The Golden Age of the American Racing Car, *a book awarded the Thomas McKean Memorial Cup of the A.A.C.A. not to mention international praise from the motoring world. He is currently working on a history of the Italian thoroughbred car. During his career, Mr. Borgeson—who has resided in Europe since 1963—has become well-acquainted with many remarkable personalities in the automobile world including Abner Doble, Wilfred Leland, Leo Goosen, Laurence Pomeroy, W.F. Bradley, Pininfarina, Vittorio Jano, Ernesto Maserati, and Roland Bugatti. He has been a fascinated admirer of Enzo Ferrari—who addresses him familiarly as* Caro Borgeson—*for years. In this chapter, Mr. Borgeson presents a probing assessment and portrait of the ever enigmatic Enzo Ferrari.*

Pininfarina was in a unique position to know and evaluate Enzo Ferrari. In his autobiography he expresses the conviction that there are only one or two really great constructors of the automobile in each generation and that Ferrari belongs to the top level of that elite. I would go much further and say that there is no real match for Enzo Ferrari in all of automotive history. The only other personality with whom he can be, and often is, compared is Ettore Bugatti, the king of high-performance motoring in the pre-World War II world. For the last quarter-century, almost, that crown has been Ferrari's. His achievements have grandly surpassed those of all predecessors and contemporaries in his chosen field. He is usually the first to acknowledge his indebtedness to others and the last to call attention to the primordial significance of his own role in this

drama which has rocked and changed the automotive world. It has been the inexorable working out of a solitary, gigantesque will, the will of an obsessed and consummate genius. In terms of one man's impact upon the thoroughbred automobile Ferrari stands alone. In fact, he stands alone on any terms.

In youth and age he has always been a splendid figure of a man—tall, powerfully built, dynamic, vital. His handsomeness is classical; given a toga and laurel wreath he would make a convincing member of the Roman Senate...or a Caesar. He speaks a rich and eloquent Italian, good French, and even says *cheese* for English-speaking photographers. He has a good sense of humor, an engaging smile, a rollicking laugh, flashing eyes. At the same time, his dignity and authority are absolute. There is only one leader in his powerful little enclave, which used to need only a moat to complete its feudal air. Until his fusion with Fiat, employees walked on eggs, hoping to avoid his wrath, and visitors risked being ego-whipped as they presented themselves at the portals of Ferrariville. He is a *very* hard man and I can't say that I have ever heard any kind words spoken of him. Awe, outrage, fury, respect, admiration, even compassion, yes. I have never heard anyone hint that Ferrari is "nice," or something of that sort. Ferrari is a self-declared mono-maniac, living for the sole purpose of building racing cars and racing them. And the goal of racing, of course, is to win, to conquer, to make one's superiority publicly manifest. Victors don't have to worry about being nice guys.

The public image which Ferrari enjoys is based upon journalism: countless millions of words have been devoted to his cars and their exploits and hardly anything to the man himself. He is in his element among journalists, being one himself, as we shall see. He speaks their language, and at the big press reunions, which he seems to delight in organizing and conducting, he commonly gives the impression of being ebulliently outgoing, open. But I know of no one who has ever gotten at all close to Ferrari. His long-time partner and admirer, Pininfarina, once spoke of his "character, closed like a walnut." This basic inaccessability, plus a thick facade of folklore, contribute to the fact that, in spite of all his fame and notoriety, very little really is known about this colossal figure.

The most important single work on Ferrari is, by far, his autobiography, *Le Mie Gioie Terribili—My Terrible Joys*—first published in 1962. It was structured in the form of answers by Ferrari to questions posed by an interviewer who was never identified. It came out having the stiff and choppy character of the transcript of a series of formal, tape-recorded interviews. The book was translated

Maranello, circa 1920.

into several languages and went through three Italian editions. Then, in 1964, came the fourth edition, bearing the sub-title *Due Anni Dopo –Two Years Later.* In this version the disturbingly unknown interviewer was disposed of, his questions replaced by smooth transitional passages which gave the work a highly readable, flowing quality which it had previously lacked. Except for this improvement and the addition of a small amount of new material—such as the ruptured romance with the Ford Motor Company—the really excellent original text remained unchanged. It revealed an Enzo Ferrari whose existence was something of a surprise: a man of broad refinement and culture, profoundly concerned with the moral meaning of his life and almost Proustian in his self-analysis. It is a really precious contribution to the literature of the automobile and, as a matter of fact, to that of psychology. Enzo Ferrari defends himself very well with the pen and, because of the intimately personal nature of much of this confessional, it is out of the question that anyone wrote it for him. How he managed to find the time to do it I do not know. But then, of course, he is no ordinary mortal.

It should be enough to keep a large organization occupied full time ''merely'' conceiving, designing, and building cars of superior merit to compete in such diverse categories as Formula One, Formula Two,

Sports, Sports Prototypes, the European Mountain Championship, and Gran Turismo. Enzo, at seventy-six, still is profoundly involved in all of this. Then there is also the activity of planning, orchestrating and directing the racing campaigns of all these machines, on a world-wide basis. Enzo says that he doesn't go to races because he can't bear to see his creations die—and even victorious cars die. But I feel that he also can't afford the time away from his telephone and teletype obligations throughout the world on any racing weekend. There are too many decisions that only he can make, or that he will let no one else make.

In addition to these duties he, personally, wages an eternal series of battles with the FIA and other controlling bodies over the rules which determine what he can build or race. He conducts another vastly complex running war with race organizers, haggling and bargaining over appearance and prize money. He picks his drivers and is a full-time talent scout. And, of course, he chooses all the men to whom he reluctantly must delegate responsibility. Of course he has usually had, for example, a financial manager. But at any hour of the day he knows precisely where his kaleidoscopic fortunes stand.

These are just some of his activities in connection with his racing empire. But then there is his GT-car empire, which exists merely to finance the other. It involves a splendid and quite large and modern factory—conceived by him. The GT business is more than enough to take all of a brilliant organizer's time and energy but Enzo keeps it, too, under his masterful control—or did until 1969, when he accepted Fiat as a partner.

When I first made contact with him in 1950 he was handling his own press and public relations personally. Of course he found good talent to help him with this, but he has never drifted very far from that sphere. He runs his big press conferences, administers an annual competition for journalists, has written a great deal for periodicals and is reputed to read everything that is published in the world press concerning himself and his cars. He is said to find the time, above all, to pay daily visits to his son's grave, where he meditates and ''works.'' One could go on and on with a list that would still not approach completion. He is the sort of genius who can play many games of three-dimensional chess simultaneously, win most of them some of the time and all of them a lot of the time.

He is unique. The name of Ferrari is one of the most common in his part of Italy. The last time I saw a Modena telephone directory there were a couple of pages of Ferraris, including one named Enzo. This competition, too, he has overcome because, regardless of the others, there is only one Ferrari who is a legend throughout the automotive world. He has the right to the titles of *Cavaliere, Commendatore,* and *Ingegnere;* he claims that he prefers simply to be called Ferrari. There is only one, after all.

In spite of his unbelievable dynamism, it must have taken a few

years of spare-time penmanship to compile his confessional. This is confirmed by another very valuable book, *The Ferrari,* by Hans Tanner. Although it was first published in 1959, its first two chapters, which are biographical, are largely verbatim translations of material which, in Enzo's book, did not see print until three years later. Tanner lived in Modena and had access to the Ferrari organization, which gives his book its great authority. It would appear that he was given an early draft of Enzo's memoirs; he may even have instigated the writing of them. What is enlightening in comparing the two versions is the passages which, it would seem, Enzo later chose to delete.

Tanner's other chief source for biographical data on Ferrari is a rare and very fine little volume by Carlo Mariani, titled *Appunti di Storia–Footnotes to History*—and published in 1957. It was published by Shell Italiana and is essentially the very exciting story of the all-important development of racing fuels by Shell from the mid-1920's—with Alfa Romeo—to the mid-1950's—with Ferrari. The book is also the story of the work done by Shell's petroleum engineer, Stefano Somazzi, and his work with Vittorio Jano, Sir Harry Ricardo, and many other giants of high-performance race-car engineering. And Enzo Ferrari is like a glowing thread running through the entire tapestry. Although Ferrari wrote the preface to the book, Mariani deals with him with a detachment which is necessarily not to be found in the two references mentioned above.*

Yet another priceless source is the already-cited Pininfarina autobiography, written by one of the few men who could deal with Ferrari on a basis of more or less absolute equality. Beyond these sources, to my knowledge, the rest of the reference paydirt lies in innumerable magazine articles and newspaper columns. Many of these were written and signed by Ferrari or suggested by him to other writers.

Modena lies in the Po River Valley about 160 long miles from Turin, as the crow flies, 105 miles southeast of Milan, and 70 miles west of the Adriatic city of Ravenna. Although that far inland, its flat and fertile plain is scarcely above sea level, fed with alluvium from the Appenines, which erupt just a few miles to the south. Modena is always seething with foreigners, not so much due to the presence of Ferrari and Maserati as to the fact that it marks the junction of the

Enzo Ferrari (left) and his brother Alfredo, circa 1904.

*Franco Gozzi, Ferrari's veteran public relations manager, was present when Hans Tanner was conducting his research. According to Gozzi, Tanner did indeed interview Enzo concerning his youth. The similarity between the Tanner and the Ferrari texts is due simply to the fact that Tanner was a very careful listener and that these were oft-told tales which Enzo told by rote.

As for *Appunti di Storia,* its author, Mariani, was a publicist working for Shell. His chief sources of information were old Shell employees, Enzo Ferrari, and Francesco Bellicardi, general manager of Weber. Most, if not all, of the book's photos came from the Ferrari archives.

1913
Ferrari at age fifteen.

Brenner Pass route to the north and the *Autostrada del Sole* to the south. It is an attractive ancient city, with a population somewhere in excess of 120,000. Its national fame derives from a gastronomic delicacy called *zampone*—stuffed pig's feet—and from a sweetish sparkling wine named Lambrusco. It is not always easy to find bad bread in Italy but the plaster-of-Paris variety typical of this region is the worst. Modena's international fame, however, is the skill of its craftsmen. Absurdly gifted artisans abound, so that you can have almost anything made, made surpassingly well, and so cheaply that you never get used to the miracle. It's an incredible place, where master pattern makers are a dime a dozen and skilled metal workers of every kind seem to surge out of the black humus. The whole Po Valley is very rich in this sort of talent, but Modena just happens to be one of its focal points. Hence the stature, in the automotive world, of this otherwise bucolic and utterly boring backwater of civilization.

Enzo Ferrari was born on the outskirts of Modena on February 18th, 1898. In his book he mentions having had an older brother named Alfredo but lets the names of his parents drop into limbo. His father was a structural-metal contractor, specializing in roofs and bridges. The modest family home was attached to the workshop and the boys lived in the room above it, being shot out of bed every morning by the first notes of the *Anvil Chorus*. Enzo recalls that his father's work force ranged from fifteen to thirty men, suggesting a certain bourgeois rank. Alfredo was an excellent student but Enzo detested school. His father's ambition was for him to become a real engineer. To his profound later regret, Enzo wanted only to drop out of school and to start earning his own living, he says. But this highly articulate man, devoted to Stendahl and d'Annunzio among many other authors, obviously was not always behind the door when the culture was being passed out.

According to the Tanner text, Enzo's father was one of the first men in his part of the world to own an automobile, and the machine "completely absorbed" young Enzo's interest. This possible myth, which imputes real wealth, is deleted from Enzo's signed version of a similar text. In it, he attributes his first exposure to the automobile to a race at Bologna, to which his father took the two boys. There, they saw the great Felice Nazzaro and Vincenzo Lancia in action. According to Enzo's story, it was not the cars that fired his imagination; it was the drivers. After having seen another race or two he had his future figured out: he would become a Grand Opera tenor, or a sporting journalist, or a racing driver.

The year 1916 was disastrous for the family. Ferrari's father died of pneumonia early in the year and some months later Alfredo died, Enzo says, from a malady acquired in the army. Without mentioning what became of his mother, Enzo says that he found himself alone and afraid in the world. He went into the army himself in 1917, soon fell

The Baracca *Cavallino Rampante*

terribly ill and was consigned to a camp for hopeless cases. He recovered, however, was discharged, and in the cold winter of 1918-19 he offered himself on Turin's glutted labor market. There he succeeded in finding a job as driver for an outfit which was converting war-surplus light trucks into passenger-car chassis, taking them to Milan for coachwork, and selling them to a public starved for wheels. Enzo had the job of driving the converted chassis from Turin to Milan. Thus began his automotive career.

As he matured, it became clear that Enzo's operatic ambitions were in vain, due to shortcomings of ear and voice. But the would-be sports writer and/or racing driver was wonderfully situated to go to work on realizing these goals. Both in Turin and Milan he made his headquarters in the bars which were the habitual hangouts of the car-racing fraternity in each city. He soon came to know more or less everyone who was anyone in Italian motor racing. In Milan he became a good friend of Ugo Sivocci, a former bicycle racer who had become chief test driver for the now long-vanished CMN marque. CMN was planning to do a little racing and Sivocci would be in charge of the team. Enzo got a job as Sivocci's assistant, which made it possible for him to drive in two events: the 1919 Targa Florio, in which he came in ninth, and the 33-mile-long Parma to Poggio di Berceto hill climb of 1920, in which he finished third. He had become a racing driver without hardly trying.

Alfa Romeo, which had abandoned automotive production in favor of heavier hardware during the war, was launching a new automotive program and jobs were available there with a more promising future than CMN could offer. Enzo moved over to Alfa as a test driver and was soon able, he says, to get Sivocci taken on as chief of that department. Following general practice in the industry, Alfa's racing drivers were drawn from the test-driver ranks. Thus Enzo got in on the ground floor of what was to become one of the greatest racing organizations in all of automotive history.

Appendix XII in Tanner's book is titled *Enzo Ferrari's Personal Racing Record* and presumably was supplied to the author by an official Ferrari source. It is dreadfully inaccurate, if one checks it against the official record book of the Automobile Club of Italy. It lists eighteen events in which Enzo is supposed to have classified, between 1920 and 1931. According to the ACI records, however, Enzo did not classify in five of the eighteen, a little error of twenty-eight percent. And there are other errors here. The correct record, according to the ACI, is shown in the accompanying box. It credits Ferrari with rides which he may have forgotten.

Ferrari gives the impression of having been quite a hotshot as a racing driver and having made a real sacrifice in giving up this career when his son Dino (Alfredo-Alfredino-Dino) was born in 1932. The "career" wasn't much, really. Aside from the Targa Florio, where his luck ran out after 1921, most of the events he took part in were bush-league affairs or hill climbs. The latter Enzo himself is the first to scorn, relative to road racing. As a member of Alfa's racing department he was in a position to get as good rides as his teammates Ascari, Campari, Sivocci, Nuvolari, Varzi, and others, *had* he possessed the talent. Instead, his personal racing record looks more like that of a dilettante or, from the mid-Twenties onward, like that of a car dealer doing a little sales promotion in and near his authorized territory. Most of his rides were in old Merosi-designed touring cars. The one good ride at which he had a chance was in a P2 reserve car in the 1924 French GP at Lyon. He was taken ill on that occasion and was unable to mix it up with the giants. It was in other areas that he was to manifest his transcendant gifts.

Between the time he joined Alfa in 1920 and certain events which, as we shall see, took place in 1923, Ferrari moved into a position of great confidence in the esteem of his employers. Just what the basis for this was has never been adequately explained, although an understanding of it is needed in order to explain how and why Alfa's vast racing activities were entrusted to Ferrari's personal generalship in the Thirties. Mariani gives us a clue in the Shell book when he speaks of Enzo's great organizational talent being recognized early during his Alfa years. Ferrari's superiors would have been blind indeed not to have perceived the embryo of genius on its way to maturity.

One executive at Alfa who was not blind to Ferrari's endowments was the youthful engineer Giorgio Rimini, the firm's sales director and therefore the head of its racing activities, the purpose of which was to promote sales. Enzo praises him for *his* outstanding organizational talents and says that Rimini "gave me the push for the formation of that complex—a veritable General Staff—which was to determine Alfa's fortunes for many years to come." Rimini sold Enzo his first car, duping him rather badly in the process while teaching him the importance of subtlety in the wording of sales contracts. Realizing that he had been the victim of his own naïveté. Ferrari took the deception as a lesson, and apparently became the protégé and understudy of Alfa's sales chief.

Concerning the "veritable General Staff," Ferrari explains that he had developed a very deep and complicated devotion to his work. Alfa's chief engineer, Giuseppe Merosi, he says, was a good man even though only a surveyor by training. His chief aide, Santoni, was a good man, too, if only a pharmacist. But these men were hardly ideally cast for epic automotive roles. Enzo says that he conceived the idea of relieving Fiat of a few of its most brilliant young technicians. He began by luring to Alfa Luigi Bazzi, a friend of his and a member of the then enormously powerful Fiat racing department. Bazzi went to work de-bugging Merosi's Type P1 racing Alfa and eventually suggested that a particularly brilliant young engineer, Vittorio Jano,

The Ferrari emblem

would be well worth stealing away from Fiat. Ferrari sought out Jano in Turin. He spoke with this total stranger, "convinced him, and he signed up the following day." And in Jano's wake, other Fiat talent migrated to Milan. Ferrari said in retrospect, "I brought to Alfa the technicians she needed. I was a protagonist, and perhaps the decisive one, of two great cycles of racing history, that of Fiat and that of Alfa...the egotistical fulfillment of my competitive yearnings."

Here, Enzo is taking the credit for driving the mighty Fiat out of racing, something which he was to do, later, to Alfa Romeo. Fiat's disenchantment with racing had fairly complex causes, which certainly included Jano's defection. They also included the defection of Bertarione to Sunbeam-Talbot-Darracq and Fiat's loss of interest in running a racing technical center for the benefit of its competitors.

Ferrari's version of the seduction of Jano is not, however, the one which Jano told me. Indeed, Ferrari made the initial contact, as Bazzi had proposed to his superiors at Alfa. Jano listened very coldly to Ferrari's pitch, then told him that if Alfa were serious they could send one or more of their top executives to negotiate with him; he was not about to give serious consideration to the proposals of a mere errand boy. Enzo made his report in Milan and top brass from Alfa courted and won Jano with a deal which he could not afford to reject. Then, and only then, did he take the fateful step. But this difference in detail does not deny Ferrari's role as an instrument of destiny in the affair. And all authorities—including experts on Fiat history—concur that, by 1923, Ferrari was an important industrial spy and pirate for Alfa Romeo.

Ferrari at the wheel of an Alfa Romeo P2 during practice in Monza in 1925.

Ferrari's greatest racing mileage was chalked up in 1923, if one assumes the full distance for the Targa Florio, which he did not finish. His one other race was that of the Savio Circuit, around Ravenna, which he won, driving a Merosi three-liter car. It was there that he also won his coat-of-arms.

As he tells the story, when he won that race at Ravenna he made the acquaintance of the parents of the young Count Francesco Baracca, Italy's World War I ace of aces, who brought down the staggering number of thirty-four enemy planes before meeting his own death. Enzo says that Baracca's mother said to him, ''Ferrari, put on your cars the prancing pony of my son. It will bring you luck.'' The parents, he says, confided the emblem to him; he added the yellow background—the color of Modena—to the rampant colt.

The first thing that I would like to say about this anecdote is that Baracca's parents were in no position to ''confide'' the emblem. It had not been their son's property—it was the insignia of the squadron in which he flew. Granted, this is a lint-picking point. But anyone who has learned the importance of contractual fine print will also respect lint.

The next point is this, and I find it fascinating. Pininfarina naturally had studied Enzo's book with intense interest. Being a family man as only an Italian patriarch can be, he was probably pained by the steamroller egotism of the Ferrari account. Although he was still Enzo's effective partner, he put into print *his* version of that anecdote: Enzo's brother—also known affectionately as Alfredino or Dino— had been a member of the same squadron as Baracca, and had met his

Left: Scuderia Ferrari at Livorno on August 3rd, 1930. (L.-R.) Borzachini, Bignami, Siena, Ferrari, Sguanci, Nuvolari, Bertolini, Augusto Caniato, Arcangelo, Ongaro, Verdelli e Lucchi. Below: Nuvolari in an Alfa Romeo Tipo B during the 1935 Coppa Acerbo.

Enzo and nine-year-old Dino. Laura Ferrari and her son in 1954.

death in the service of that squadron. Hence Baracca's mother suggested that Enzo would do well to adopt the insignia of his own brother's squadron, and that of her son. And, finally, one learns from the official biography of Franceso Baracca that he was a graduate of the military academy of. . .the city of Modena.

Jano knew all there was to know about the 1923 Grand Prix Fiat, including its defects and how they could be corrected and improved. He designed a Fiat-killer in record time, aided by the iron discipline which he imposed and by the appetite for it shared by the whole "General Staff" which orbited around this messianic engineer. The resulting car, the Type P2 Alfa Romeo, was a world-beater from its début in the 1924 French GP. It routed Fiat and all other contenders in top-level international racing and it launched Alfa Romeo as an international racing marque of the first rank.

Despite the P2's continuing invincibility and winning of the World Championship, affairs were going very badly for Alfa Romeo. Italy never had been very strong economically and, although on the "winning" side, never could recoup the disastrous cost of participation in World War I. A government organization called the *Istituto di Ricostruzione Industriale* was created to take over bankrupt enterprises which were important to the national economy and keep them producing; the IRI still exists and is a very important force. It absorbed Alfa Romeo in 1925 and remains its owner today.

Following the death of Antonio Ascari in a P2 during the French GP at Montlhéry on 26 July 1925, Alfa Romeo withdrew from racing on the pretext of mourning for its fallen champion. The real facts probably were more directly related to economic reality. There was no money left in the company coffers, the marque's image had been glorified by racing triumphs, and the time had come, dammit, to sell a few cars. Racing on the taxpayers' money was quite out of the question, and Alfa's stable of racing vehicles was the first to be sold. . .to discriminating sportsmen. This did not mean that the IRI was at all unhappy with the sporting image which the marque had been cultivating. On the contrary, Jano was given the go-ahead on a crash program for a line of new and *really* sporting, thoroughbred vehicles which the already race-conditioned public could buy. Enzo's activities during this difficult period are another mystery, but I would guess that (1) he was very useful in the new sports-car project and (2) he functioned as factory go-between and consultant to that very special clientéle which had purchased and was amusing itself with the Alfa racing stable. Mariani tells us how Enzo was constantly increasing his sphere of contacts in the racing world. He got to know almost everybody in Italian motor racing, including the hungry bounty hunters and the well-heeled sportsmen-dilettantes of the so called Gentleman Class. When he became an outright car merchant is uncertain, but it was no later than December 1st, 1929 when, with a handful of share-holding partners, he founded the Scuderia Ferrari in Modena. It held the exclusive Alfa Romeo franchise for the local province of Emilia and for the adjoining provinces of Romagna and Marche — a lovely, large territory. But if one reads official Alfa historian Fusi correctly, Enzo was selling cars for Alfa Romeo well before the Scuderia was formed.

Alfa, which had been officially absent from racing since 1925, returned to the fray with the Mille Miglia of 1928, then the Targa Florio. Enzo was not entered in either event. This resumption of racing, with the new Jano-designed sports cars, was perfectly timed to coincide with the advent of the Great Depression, and Alfa was soon constrained to drop official racing once again.

The word *scuderia* means stable and, by extension, racing team. Part or all of the initial concept behind the Scuderia Ferrari seems to have been dual: (1) to sell cars to all comers and (2) to operate as a club and service center for its racing members. Whether it was foreseen as a smokescreen for Alfa's continuing participation, as a factory, in sports-car and Grand Prix racing is one of the many important points which is thoroughly unclear in all references to the Scuderia. Mariani regards Ferrari's personal racing career as having effectively ended in 1928 but stresses his great and broad involvement with Alfa's racing clientéle. It was so complete, he says, that when the Scuderia became a front for Alfa's racing activities, this was universally regarded as natural working out of a logical process.

In his book, Tanner almost gives the impression that Alfa became a subsidiary of the Scuderia. Ferrari says that he never went off the Alfa payroll and that the Scuderia, while "independent" of the factory, nevertheless was "umbilically tied" to it. Fusi says that Enzo's title within the Alfa organization at this time was Sales and Racing Consultant. In any case, the Alfa racing department went underground, and its racing cars and drivers resurfaced in Modena in 1930 under the sign of the rampant colt. A number of technicians were transferred there, plus much important tooling, including a good engine dynamometer. Jano and Somazzi became regular visitors, and it seems that from the outset of the 1930 season Alfa's racing department was snugly installed behind its new façade.

What has come down to us concerning the history of the Scuderia Ferrari is maddeningly hazy, superficial, fragmentary, evasive. This is most unfortunate since it is one of the most significant, exciting, and action-charged episodes in automotive history. And it is the bridge which joins two great racing traditions—those of Alfa Romeo and Ferrari. Many people are still living who lived this experience at first hand and their knowledge should be recorded while there is still time. And the story of Enzo's years in Milan also should be excavated with an archaeologist's care. Only when we know many of these missing

Relaxing: Bonetto, Ferrari, Giberti and Villoresi.   **Villoresi and Ferrari in 1952.**   21

details will a coherent record of this epic be within our reach. What is self-evident, however, is that the Scuderia functioned marvelously well and that when Alfa entrusted this staggering challenge and responsibility to Enzo Ferrari it shrewdly chose one of the few men in the world—and perhaps the only one—capable of carrying out the job. And, of course, it was a fantastic school for Ferrari, the crucible in which his genius was refined, the legacy of unique experience on which his future conquests would draw.

From the outset, one of the Scuderia's functions was to provide feedback from racing to the factory's experimental department. This activity began on a modest scale but became increasingly important, until entire racing cars were being designed and built, almost in their entirety, in Modena. As Ferrari sees it, *he* was the creative and directive force behind these projects, and the instigator of great decisions. The Alfa Romeo point of view stresses decisions made in Milan and carried out in Modena by factory technicians who were assigned to work in that ivory tower of motor racing. Fusi gives us the official list of these projects:

1933—Boring out the Monza engine from 2336 to 2556 cc.

1935—Boring out the P3 engine from 2905 to 3165 cc, installation of Dubonnet independent front suspension and cantilever rear springs on the P3; construction of the Bimotore, powered by P3 engines fore and aft.

1937—Construction of the Tipo 158, many of the parts of which were made at the factory in Milan.

1938—Construction of the Tipo 308, another monoposto, based on 8C 2800A GT components; construction of the 312 monoposto, out of 12C 1937 components; construction of the 316 monoposto.

This feverish activity in 1938 reflected Alfa's all-out effort to counter the challenge, in motor racing and in technological sophistication, of Germany, into whose camp Italy had been thrust by the Western Allies. One of the consequences of this effort was the dismantling of the Scuderia Ferrari façade and the move back to Milan of the entire Alfa Romeo racing organization, under the auspices of a new subsidiary named Alfa Corse. Along with the rest of the personnel and hardware, Enzo was shunted back to the factory. As he recalls it, nothing changed at all except his title, which became Director of the new racing section. According to Fusi and the factory records, it became Directive Consultant, which is slightly different.

The Scuderia years had been an extraordinary experience for Ferrari. They taught him an infinity of things including, he says, how to synchronize a commercial sales and service organization with the activities of one of the world's best and most active racing organizations. He got to know "men and things" as he never had dreamed of knowing them: race organizers, superstar drivers, company

An early 625 in 1954.

executives, and just how the whole business really worked.

One of his most prized memories of that period is when "a project was born for the realization of a racing car that was entirely 'mine.' Thus was born at Modena the Tipo 158 Alfa...with which the Milanese house was able to win two World Championships...It was born in Modena in 1937, in the Scuderia Ferrari, from one of my personal ideas and realized through my will."

This is a good example of Enzo's sublime arrogance and of his limitations as a reasonably objective witness. Alfa Romeo had been crushed in Grand Prix competition by the three-liter Mercedes-Benz and Auto Union machines. All that was left as a hopeful arena for Alfa was the 1500-cc Voiturette category, then the Little League for one-man racing cars. The Alfa 158 was inspired by the brilliant little supercharged 1500-cc Maseratis, the cars to beat. Four 158 Alfettas were indeed designed at Modena by Alfa engineer Gioacchino Colombo and assembled there, from parts made on the spot and in Milan. Mariani, who misses no opportunity to assign credit to Ferrari, ignores him in this case, as does Fusi. But this doesn't matter since, before it could begin to realize its potential, the 158 had to be drastically redesigned in Milan by Ing. Orazio Satta and his staff. It was the resulting Tipo 159—worlds removed from the 158—which became the phenomenal world-beater. Jano, meanwhile, considered both to be mere variations on the traditional theme which he had steadily refined since the P2 of 1924. Exaggeration is superfluous in appraising the historic significance of Scuderia Ferrari.

Enzo's recall to Milan was marked by the signing of a new employment contract. It specified that, if and when he should leave the company for any reason, he would have to stay away from car racing and racing cars for at least four years. For eight years he had been the big frog in Modena's little pond; life was not the same back in Milan. He says that he began to make felt "with instinctive arrogance, my vocation as agitator of men and of technical problems. I have to say that what I was then, I am now; never have I thought of myself as a designer or inventor—just an agitator."

Back under Alfa's roof, his old freedom to "agitate" was severely whittled down. He was under the serious, watchful eye of Alfa's managing director, Ing. Ugo Gobbato, an industrial leader for whom the most precise possible long-range planning was a religion. Enzo was at least as religious, but committed to the worship of spontaneity, immediate necessity, and the omnipresent unforeseen. Each could only regard the other as aberrant. Then there was the presence on the Alfa Corse staff of the Spanish engineer, Wilfredo Ricart. Many of the men who had to work with and under Ricart—who later created the Pegaso automobile—could support neither his personality nor his engineering ideas. Enzo felt the same way, and later events were to

23

vindicate the anti-Ricart faction. But Enzo had the responsibility of transmitting all of this criticism to Ing. Gobbato, who happened to be an admirer of Ricart. He accused Ferrari of being envious of *lo Spagnolo* and it seems to have been in November of 1938 that Enzo walked out of the Alfa factory gates for the last time. It was a painful experience, but Enzo consoled himself with the counsel that it was all for the best: a man should not stay in one spot too long, and this had been a very long stay.

Scuderia Ferrari had been liquidated and Enzo now had a good chunk of capital of his very own. He still owned the old Scuderia building in Modena and there he set up the Società Auto Avio Costruzione, doing sub-contract machine work for a manufacturer of engines for trainer aircraft. But he lost no time in getting back into the car-racing game. With the collaboration of Ing. Alberto Massimino and of Enrico Nardi, he developed a 1500-cc competition two-seater, largely based on Fiat components. The project began in the winter of 1939, and two cars were ready for the last prewar Mille Miglia the following April. One was driven in that race by young Alberto Ascari and the other by the Modenese aviator, the Marchese Lotario Rangoni Machiavelli. Neither car survived the test, and one went back to Turin to become the once famous Nardi-Danese. Because of his contract with Alfa, these "first Ferraris" were known only as the 815 and were not associated with Enzo's name.

At this same period, Nardi was to play a much more important role in Ferrari's future. He introduced him to an important Torinese dealer in machine tools who made Ferrari aware of the existence of a gaping market for good domestically-built copies of German oleodynamic grinding machines. Ferrari acted on this lead and did so well at Modena, where he had the space for about forty workers, that later in 1943, he was able to build a plant on the outskirts of the nearby village of Maranello which could accommodate four times that number. He seems to have passed the war years quite profitably, in spite of a couple of bombings of his plant. As soon as the war was over he phased out of machine tools and re-set his sights on racing cars. Of course, he had been thinking of nothing else for years and he had shared his hopes with his old friend Gioacchino Colombo, of Alfetta 158 fame.

Colombo had gone to work as a draftsman in Alfa's engineering department in 1924, mere days before his twenty-first birthday and very shortly after Jano's arrival there. Both he and Fusi had eventually become top aides to Jano and expert exponents of their maestro's mechanical ideas. Colombo had been sent by Alfa's management to Modena in May of 1937 where, in the closest contact with Ferrari, he had worked successively on the designs of the Alfa 158, 308, 312, and 316 Grand Prix cars.

Political life was particularly difficult in Italy between the wars.

Enzo had received his titles of Cavaliere and Commendatore from the Fascist government for some obscure reasons. That they did not involve political complicity seems to be confirmed by the fact that even Enzo's critics and enemies do not accuse him of that sort of collaboration. Colombo, on the other hand, was prominent in Party politics and therefore found himself to be a pariah when peace came and the Resistance inherited all power in Italy, including the power at Alfa Romeo. Now, with the war ended, Ferrari, not exactly the man to blame anyone for loyalty to iron dictatorship, gave Gioacchino Colombo the green light to design a new racing machine. The result, of course, was the original 1500 cc Ferrari V-12.

Explaining this architectural choice Ferrari says, "The twelve-cylinder is an engine that I have always cherished, remembering the first photos of a twelve-cylinder Packard, which raced at Indianapolis in 1914" (sic). Then he tells of observing the Packard Twin-Sixes of top American Army officers immediately after the First World War, and loving "the harmonious voice of this engine." He offers these whimsical thoughts in justification for a daring engineering decision which, he says, made him a laughing stock until his cars began winning everything in sight.

This is yet another little tale which, as it is presented, merits no respect. It could mislead European readers, impressed by multiplicity of cylinders and knowing absolutely nothing of the performance potential of the primordial Packard Twelves. There may be a very fertile clue in this apparent nonsense, but it would be a major project to explore its possibilities. But the possible clue does not matter here. What is significant is a deliberate effort to distract the reader from the real sources of the original Ferrari V-12 and to invest Enzo with its immaculate conception, for sentimental reasons.

As usual, Enzo was drawing heavily on his Alfa Romeo education, as was Colombo. In 1931 Jano designed the Alfa Romeo Tipo A V-12 monoposto. He followed this in 1936 with another V-12, the Tipo C. Next came the 12C 1937, then the 312 of 1939. There also was the V-12 S10 Prototype GT of 1938. Then the V-12 sports-racing Tipo V-12 of 1939. Alfa Romeo had developed magnificent full-race V-12's throughout the Thirties, and both Ferrari and Colombo were saturated with knowledge of them. If that were not enough, just before the war the Germans had pointed the way by crushing all opposition with V-12 engines. It was the way to go at that time, and it was the way that Ferrari went, although it was a costly way.

When the young engineering draftsman Aurelio Lampredi first went to work for Ferrari in September of 1946 there seems to have been no V-12 program. Dissatisfied at Ferrari's, he left in February of 1947. That November, Ferrari enticed Lampredi back to Modena and showed him the general layouts, which Colombo seems to have prepared in the interim, for what was to become the Tipo 125, a 1500 cc V-12 and

World Champion Fangio and his Lancia/Ferrari.

Fire at the Nürburgring, Scarlatti escapes.

Above: Dinos at the Nürburgring.
Below: Hawthorn's car for the 1958 Monza 500.

the ancestor of the entire Ferrari V-12 line.

Colombo came from Milan to Modena only about once a week; would Lampredi consider taking over as the full-time draftsman on the job? He accepted, and Colombo was seen more and more rarely until he stopped coming to Modena entirely in 1949. Ferrari has remained very loyal to Colombo, while having little to say concerning Lampredi. But it was Lampredi who de-bugged the 125, as Satta had de-bugged the 158. The absolutely unknown Lampredi went on to one prodigious design after another for Ferrari, creating much of the patrimony of the marque and insuring its early racing success. But it is doubtful that Lampredi ever was appreciated at Modena, and when his very close friend, Alberto Ascari, lost his life in a practice crash at Monza in 1955, Lampredi flatly quit. He became the head of Fiat's engine-engineering department and the apparent father of the subsequent family of Fiat dohc powerplants— ''Alfas for the mass market.''

Ferrari, of course, found a replacement for Lampredi and continued to build up an ever-stronger technical and executive staff. He continued to flourish until the end of the 1961 season, when his key personnel walked out *en masse*. This cataclysm was reported in British *Motor,* in an editorial piece which referred to ''Enzo Ferrari, the egotistical genius who has never been able to abide even constructive criticism.''

The exodus consisted of ''Girolamo Gardini, for years commercial manager and widely regarded as 'the man behind Ferrari'; Carlo Chiti, the chief engineer; designer Giotto Bizzarini; Romolo Tavoni, team manager; Ermano Dellacasa, the financial expert; Federico Giberti, factory manager at Maranello; Enzo Selmi, personnel manager and Giorgio Galassi, responsible for running the unique Ferrari foundry.'' Top driver Phil Hill also quit.

''The reason for the walkout,'' *Motor* continued, ''has never been made clear. It was asserted (and denied) that interference on the part of millionairess (in her own right) Signora Laura Ferrari was the root cause. Enzo Ferrari never watches his cars race, and for a few seasons his wife represented him on the circuits, telephoning regular reports to her husband. Were those secret reports the reason for the breakup of the world-famous motor-racing team?''

Ferrari had been Destiny's darling up to this point, a great talent acting as a magnet for great talent throughout the world. He had worked harder, perhaps, than a thousand normal men. Yet the most crucially important things seemed to arrange themselves effortlessly, such as Lampredi arriving out of nowhere and catapulting him into a position of world leadership. Obviously, talent was cheap and could be replaced at will. What counted was Leadership. The ancient Greeks called this disease hubris, the poison of pride.

Ferrari has had a fine eye for talent but he has been plagued by an

inability to hold onto it. This very grave fundamental problem has troubled Enzo's relationships with drivers throughout his career.

He criticises them for their lack of loyalty to a single house. Yet half the good drivers in the world have entered his house and found it inhospitable, the host devoid of any positive feeling, including loyalty, toward them. Many of them complain of his tyranny, his calculated harassment, and of efforts to psych them out, break their spirits, humiliate them, frustrate them, drive them frantic. Many feel that all that counts for him is a Ferrari victory, and that when oceans of publicity and praise are lavished on one of his drivers, he resents it bitterly and takes this out against the driver.

There is probably a good deal of truth in this and, if so, a basis for it that is not illogical. In the old days, from Cagno to Campari, there were no prima donna drivers. Even the great ones were on factory payrolls, working hard at modest jobs and justifying their keep until called upon to defend the house's colors. The rewards they received were minimal, even in terms of glory, the bulk of the credit in the case of victory going to the marque. The basic justification for the expensive folly of indulging in racing was, for the manufacturer, sales promotion rather than improvement of the breed. There were good drivers and mediocre ones, but they all knew their place relative to the marque, to the mother house.

Ferrari, raised in this ethic, met a new breed of racing driver when management of the Alfa team passed to his Scuderia. In Tanner's book we read, ''Enzo Ferrari had a problem with his team, something which would arise many more times in the future of his racing activities. He had too many prima donnas...'' In other words, too many drivers who put their own interests above those of team discipline. The worst

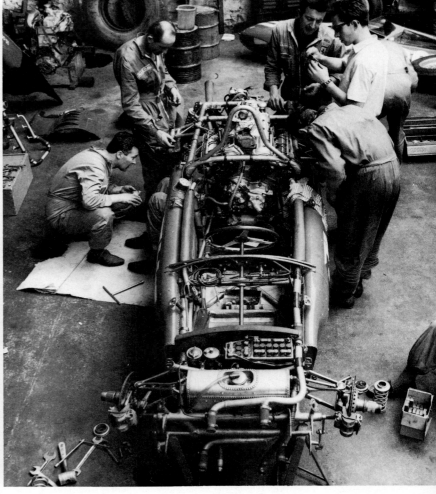

Race car preparation
at Maranello.

of these, and the most uncontrollable, were Nuvolari and Varzi, whose
raison d'être was to beat each other, and to hell with the marque.
Tanner goes on to tell how, in the Mille Miglia of 1930, Ferrari
deliberately lied to Varzi concerning his position in the race, which
enabled Nuvolari to win. Varzi stayed around long enough to trounce
Nuvolari in the Targa Florio, and then became the first of the long list
of Ferrari deserters, going directly to Bugatti.

Throughout the Scuderia years Enzo had to live with this problem,
and he developed his own methods for dealing with it. Also, having

lived in the shadow of top-rank drivers at Alfa, he may have had a
spontaneous resentment of such talents, regarding them all as prima
donnas until possibly proved otherwise. He also acquired unique
experience in the handling of the all-too-independent owner-driver
type, including the very upper-class and wealthy, who would try to get
their way by pulling rank on the king of the Scuderia. It was an
incredible school, in which Enzo had to try to impose some sense of
order and discipline upon veritable herds of delirious *Alfisti* who
would pack the Mille Miglia and Targa Florio year after year. It may

31

have left him scarred for life.

When, in 1951, Froilan Gonzalez in a Ferrari whipped the Alfettas for the first time, Enzo recalls, "I wept for joy, but mixed with the tears of happiness were those of sorrow because I thought, 'I have slain my mother.'" He wrote to Alfa's managing director, saying, "I still feel for our Alfa the adolescent tenderness of first love, the immaculate affection for the mamma." Success followed success and then, toward the end of the 1953 season, Ferrari made the first of his familiar threats to quit racing.

He was making his living by racing, which helps to explain the fantastic hyperactivity of the marque since its inception. Aside from its clashes with Alfa Romeo in Formula One, Ferrari had an effective monopoly in international racing. He was therefore able to demand, and get, very substantial starting and prize money from race organizers, plus assorted subsidies and rewards from various sponsors. He was on Easy Street. Then, with Lancia, Maserati, and Mercedes-Benz all moving in on the action, and with Alfa intending at least to continue in sports car competition, the joyride was coming to a swift end. The racing largesse was going to have to be shared with a number of others, and Ferrari was threatened with extinction.

In spite of his successful fling at machine-tool manufacture during the war, Ferrari says that he has never thought of himself as an industrialist, but as "just a builder of things and a promoter of innovations." And as for what really counts for him in life, he describes it as "...that dominant passion of life, which has deprived me of the time and the taste for almost anything else. I have no other interest than the racing machine."

The racing machine and *not* the fast luxury car, industrially built in series production.

"The productive development of my business," he continues, "can be of interest to me only if it is run by others. One solitary thing drives me—to be able to dedicate myself totally to prototypes and to racing cars, but under conditions of absolute independence..."

The life-saving solution to these problems of finance and creative energy was devised by Enzo himself and was heralded at the Brussels Salon of 1953 with the presentation of a Ferrari 212, beautifully bodied by Pininfarina.

The Gran Turismo cars which Ferrari had built up to this point were essentially sports-racing chassis equipped with bodies—usually quite stark—by any one of a variety of coachbuilders. Most of the few which were built were one-off machines ordered by individual sporting clients, and therefore very costly. There was a seller's market for fine GT cars conceived for the non-racing sporting clientele and produced, relatively economically, in series. Alfa Romeo, like Bugatti, had perfected this type of operation, and Enzo knew, from the inside,

Le Mans winner 1962, 4-liter prototype.

exactly how it should be structured. And he also knew how it could be synchronized to support a world-dominating racing organization, as Alfa had done. He knew that the staggering reputation and achievements of Alfa Romeo had been made possible by absurdly low production figures. In the two decades, 1920-1939, the marque's average production was a mere 473 cars per year. In 1936 the firm built only ten units. Only once did it break the 1000 mark, in 1925, with a total of 1115 bare chassis.

Alfa Romeo's automotive activity—as opposed to trucks, aero engines, etc.—consisted of building a marque image through racing, then selling small numbers of sports and GT cars to a highly discriminating segment of the public at quite stiff prices. The markup and profit on limited-production luxury goods must, of necessity, be substantial. Thus a manufacturer's profit of $2,000 on a $10,000 list-price car is quite reasonable. If one can sell only 100 units per year one is already making an annual profit of $200,000. If one can manage to sell 1,000, the annual profit is, of course, $2 million, quite enough to finance most expensive hobbies. Even a bad year of only 250 sales would still yield a half million dollars. The problem for the hobbyist or *appassionato* is to find someone able and willing to undertake the drudgery.

Enzo must have seen it all coming in 1951, when he opened negotiations with Pininfarina, who was still a rather small operator but whose outstanding talent Ferrari obviously recognized. Pinin describes this bourgeois courtship most entertainingly. Ferrari sent word to Turin through an emissary that he would like to have a word with Pinin in far-off Modena. Pinin's response to the command performance was to tell the intermediary that he would be happy to meet with Signor Ferrari . . . in Turin. This game seemed to go on interminably, with the two men avoiding each other, Pinin says, like two fish in a bowl. Neither would give in. Finally, ''a Solomonic solution was found: we would meet on neutral ground, in Tortona.''

Pinin won this round, with Ferrari having to travel the longest distance by far. But he did not win the next one, when he tried to persuade Ferrari to come and see his very fine factory. Ferrari made it clear that he went nowhere to see anyone, period, end of discussion. Pinin accepted Ferrari's character as he found it, and the courtship was concluded successfully. They got down to pure business and parted, understanding each other well, without so much as a drink to seal their accord.

Pinin says that the talk in the trade ran, ''Ferrari and Pinin? It will never work; it's like putting two prima donnas in the same opera or two priests in the same parish.'' Here, Pinin was acknowledging his own well-known hard-headedness. But, to the astonishment of all, it turned out to be a marriage made in heaven, each obstinate man having total confidence in and respect for the other. They complemented

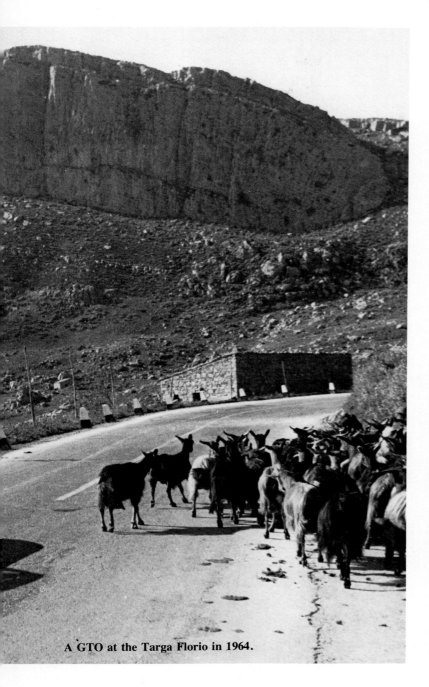

A GTO at the Targa Florio in 1964.

each other, Pinin says, "like a pair of clasped hands."

This wedding resulted in the creation of Ferrari's GT-car division, committed to the production of no more than 1,000 cars per year. Pininfarina becoming an effective partner in the affair. Thus Ferrari was able to finance the one thing in life which he claimed to care anything about.

There was something else that was dear to him, beyond even his own eloquence to express: his son, Dino. This precious gift from his wife did not alter his feelings about marriage. He states that marriage reduces a man's liberty by at least half and that no one with a passion such as his should marry. That he could have made such an error was the fault of his youth at the time. He recalls it as having been "around 1920" and does not even or ever mention Laura's name in the early editions of his autobiography.

"The conviction has stayed with me," he says, "that when a man says to a woman I love you he means to say, in reality, 'I want you.' And that the sole total love possible upon this earth be that of a father for a son."

Speaking of his relationship with Pininfarina he says, "And it was clear, at a certain point, that one sought a beautiful and famous woman to dress, and the other a *couturier* of world rank to dress her." In other words, the important "women" in Ferrari's life were the cars which he created.

It is painful to read Ferrari's tortured laments for his lost son, a theme which long obsessed him and dominated his daily life. In addition to his faithful daily visits to Dino's grave, he is known to abandon important visitors in his office when he feels the need to pray in Dino's chapel, which adjoins it. His paternal love/want/need is overwhelming, or certainly was so for many years.

He had created this incredible patrimony and tradition out of his own physical and spiritual resources. His son was groomed in every way to carry it on and to give his father that measure of immortality. Dino's death, at the age of twenty-four in 1956, was the death of the Ferrari dynasty, leaving Enzo feeling orphaned, yet somehow the spiritual heir of his son. One can feel with him when he writes, "Few men have, like Pininfarina, been able to pass on their own achievements, their own factories, their own names, to such worthy descendants. And these, in turn, have their offspring: the tradition is secure."

As it has worked out, after reportedly exploring the possibilities with General Motors, Chrysler, Ford, Alfa Romeo, and Lancia, Ferrari in 1969 handed to Fiat the responsibility for carrying on his tradition. It was an ideal choice, and he retains the presidency of the marvelous organization which he created.

The infusion of Fiat influence at Modena and Maranello has

wrought little change. The plant at Maranello and the offices in Modena, as conceived by Ferrari, were and are models of clean, modern efficiency. But relations with clients and the press have undergone a certain dramatic improvement. Not all of the old tradition merited being carried on.

In describing the original quarters of Scuderia Ferrari in 1930, Hans Tanner refers to "the notorious waiting rooms where millionaires and future racing champions were to await the pleasure of Enzo Ferrari." Mariani refers to a solitary "waiting cubicle."

It was in 1963 that Jasmine and I received John and Elaine Bond, the publishers of *Road & Track,* at our home in Turin. John needed a replacement for the broken crankshaft in one of his Ferraris, we were going to be doing a bit of touring together, and I made an appointment for Mr. Bond to discuss his problem at Maranello. I said nothing about his being shown any special courtesy, considering this to be both boorish and superfluous. After all, what *R&T* had done, out of pure enthusiasm, for the Ferrari name and sales over about a quarter-century could hardly have missed the attention of the people we were about to visit. In my capacity as working journalist I had always been received with tolerable deference chez Ferrari and was quite unprepared for what greeted us.

The four of us presented ourselves to the armed guard at the gate of the fortress-factory. He obtained instructions by telephone, told us harshly that Mr. Bond and I might enter but that the ladies would have to wait outside. Rather than sit in the car on the edge of the country road they decided to do their waiting in the only other sign of civilization in the rural surroundings—an old farmhouse-turned-bar, planted in a field across the road. After a half hour or so Elaine laughed and said, "I can't wait to tell Phil Hill about this when we get back. He's told us about having to hang out in this place when the gate was slammed on him."

Meanwhile, John and I were not greeted by anyone we knew but by a total stranger who coldly took down the data on the crankshaft and vanished, saying that he would return in a moment.

We waited in the bare cell for fifteen minutes, thirty, forty-five. Finally I asked John, "Can't you get the thing made in L.A.?"

"Yes," he said, "but I would have preferred to get it here."

"And what do you think now?" I asked.

"Let's hit the road," he said.

We parted. The armed guard disarmed the gate and gave us our liberty.

It was a pretty grim experience but I chalked it up to some honest error on the part of Ferrari personnel. Then I was invited to Maranello for the introduction of the 330 GT 2+2.

I made the dull train trip with a very close friend, the late Giorgio Colombo, the right-hand man of Pininfarina himself, the creator of the

Below: during testing at Modena: Ferrari, Tavoni, Chiti.   **Center: Enzo and Laura at Monza.**

coachwork of the car we were about to see. We arrived at the familiar gate at about ten in the wintry morning. There was the usual check by 'phone, and we were told that *we* would have to wait outside. We stood there in the thick snow as the French industrialist/racing driver Jean Guichet, of Le Mans achievements, arrived. He was told that he could not enter. "I'm here on urgent business, concerning the delivery of a racing car," he explained to the indifferent guard. He was let in, but Giorgio and I were left out and sought shelter in the *trattoria* across the road. I wondered if Ferrari owned it; if not, he should buy it. He was sending an awful lot of trade that way. I told Giorgio that I found the situation pretty insulting.

"Don't take it so hard," he said. "It's perfectly normal. My president, Pininfarina, is the only person I know of who doesn't get this sort of treatment. It happens to his son, Sergio, frequently. Be glad you're not royalty. You should know what's happened here to Leopold of Belgium and his Queen."

When the Word finally came down to let us in we had only waited for about an hour. Guichet, who was about to win Le Mans for Ferrari, had not gotten beyond the receptionist, and was livid. What his fate ever was, I don't know. When Enzo finally received us it was, "Ah! you made it! Did you have a nice trip?"

"...where millionaires and future champions await the pleasure of.."

Ferrari's ambition to become a sporting journalist was first realized in 1914 as a correspondent of the national daily, *La Gazzetta dello Sport*. In spite of the fantastic demands upon his time in the old Scuderia days, when he was still working in his own pits, he nevertheless found it possible to occupy himself with the publication and editing of a pretty interesting twice-monthly magazine titled *La Scuderia Ferrari*. It was devoted to documenting the dizzying activities of the Scuderia and to the achievements of the Fascist State. In addition to this, Enzo also got together and published quite massive yearbooks for the Scuderia. The title page of the 1933 edition pictures ex-journalist Mussolini on a motorcycle and also at the wheel of an Alfa monoposto. In 1949 Enzo published the first all-Ferrari yearbook, the first of the famous series. In the 1960's he found the time to write weekly columns for *Auto Sprint* of Bologna. He is a natural, good, and prolific writer.

He seems to revel in hosting the cream of the world's motoring journalists; he seems to feel a certain identity with them. But he knows how dangerous they can be. He fears their capacity for espionage, either deliberate or inadvertent. Their job is to report what is new and different and it is precisely this which determines where the power lies in the racing game. The most well-meaning of journalists, doing an innocent and routine job, may note and give away a secret of the utmost logistic value. No one understands this

The author and his subject.

better than Ferrari, nor is more paranoid about it.

Until the Fiat fusion in 1969—when a press conference was called for the introduction of new racing cars—no one knew what the press would get to see until Enzo had made up his mind, which could be at Hour Zero, minus a few seconds. Journalists from all over the world would arrive at Maranello, at great expense and effort, in response to his invitation. They would ask his highly placed minions what they were going to be able to look at and photograph. The minions knew as little about it as they did. You might get to see nothing but a press release and you might be given the free run of the entire plant. You could be shown the actual racing cars; then some reporter might pull out a steel tape and the party would be abruptly finished: "All right, everybody. We're going elsewhere now." But there was remarkably little grumbling about this treatment. One understood what was at stake. Everyone knew that Ferrari understood his own affairs best and that he was the King on his own turf. He did his best to help us to help him and he knew—that which we could not—exactly where the limits were. This sort of problem has been greatly eased in the post-Fiat period.

In an article titled *Ferrari Is a Personage Out of Greek Tragedy* by Sejero Baschi, Ferrari is quoted as telling a long-time friend: "I have always been certain of one sole thing: from the day of my birth someone has accompanied me on my road, constantly—Death. Actually, the thought of death liberates me from all trouble and distress. When I feel myself overwhelmed by worldly problems which I seem to be incapable of resolving, I take recourse in that sort of inevitable witness which has been at my side since I first came into the light of day, and I feel comforted, augmented, detatched."

This concept of death as a personal companion and benefactor is quite unusual, particularly in the realm of blood sports, where the merest mention of death tends to be highly taboo. Ferrari's concept may be a key to his career, talent, success, and way of life. Being frequently accused of mass murder, of dripping with the blood of a host of victims, has caused Ferrari much reflection. And others too. Phil Hill has said in a 1970 *Motor Trend* interview:

"I found myself continually wondering what are Ferrari's feelings in this involvement? What does he consider the ultimate goal, retiring after having been something super, or perhaps dying in a racing car or...I never did get an answer...A lot of Ferrari drivers had unfortunate accidents and died. I don't really understand why so many of them did, whether there was something peculiar about the kind of person that would drive for Ferrari, that set himself up for that sort of thing, or who wasn't too bright about figuring out his own capabilities."

In his book Ferrari says: "To him who comes after me I confide a

very simple heritage: keep alive that will to progress pursued in the past even with the sacrifice of the most noble human lives.''

Ideas like this are explosive anywhere, but above all in that most Catholic of countries, Italy. There, Ferrari has been repeatedly under violent attack as a murderer from many sectors, including the lay and Church press. In 1953, when de Portago in the Mille Miglia (the last) killed himself and about nine spectators, the whole country's wrath and outrage was directed against Ferrari. He did not see why he should be held responsible for what was, in his mind, merely a question of a lacerated tire. He was ready to quit racing, but a flood of mail from all over the world assured him that he had done no wrong and healed his wounds.

The final chapter of Ferrari's autobiography (fourth edition) is devoted largely to his concern for the question of his personal moral responsibility in the conduct of his career. It is not amusing, living in Italy, and having the Vatican press label you ''a modernized Saturn, devouring his own sons.'' Ferrari fills pages with arguments to the effect that great sacrifice is justified when technological—and therefore competitive—supremacy is the goal. He concludes with: ''It does not seem to me that I have ever committed a bad act. Yes, I am calm, even if not serene, even terribly imperfect, as I am. I have never repented; I have often regretted, but repented never. Is this a good thing? I fear not.

''I feel myself to be alone, after so many delirious events, and almost guilty for having survived. And how detached, because in the aridness of this earth the plant of hope can grow only if it is nourished by the love of a son.''

Whatever one's personal feelings may be concerning Ferrari's moral responsibility for the inevitable consequences of building racing cars and campaigning them, there is something less than fair in his singling out for flagellation by moralists of other persuasions than his own. He never forced anyone to drive for him, and anyone who chooses to drive a racing car knows what the chances are. And, much to Ferrari's credit, his cars have always been noted for that solidity which contributes to driver safety. It is hardly fair that he be hounded like a criminal when the targets of Ralph Nader's inquests suffer no stigma at all.

It hardly seems to be fair, too, that Ferrari never has been formally honored by a postwar Italian government...and there have been thirty-seven of them at this writing. But *everyone* knows that, from the late Forties onward, a very important portion of the prestige enjoyed internationally by Italian technology and industry is due to the single-handed and single-minded battle of Enzo Ferrari.

For all that, in his own chosen field of combat and self-assertion, Ferrari has won immortality...or has forged it for himself. He has been and will be challenged, but nothing ever can dislodge his crown. He is, very simply, The King, and the personification of mastery over a merciless discipline.

When, in 1964 and shortly before his death, Vittorio Jano recounted to me the story of his life, he had known Enzo Ferrari for forty-one years. Jano was still active as an engineering consultant to his nearly lifelong friend. He told me:

"When Ferrari sought me out in 1923 he was Alfa's sales representative for the Province of Emilia, but he was much more than merely that. He had the faith of Senator Romeo and of Alfa's other top men. He knew a little about everything and Romeo probably brought him into the company for the commercial help that he could give.

"One of the reasons for the excellence of Ferrari's machines is that he has the final word before any new model is released. When his chief test driver has finished shaking a new car down and pronounced it *a posto,* it is then Ferrari's turn to see if he agrees. In doing this he does what I used to do at Alfa. And it is here that his talent as a driver really manifests itself. He is a terrific driver at speed; I still make such runs with him occasionally, and they're really impressive. Comparing him with Alfa's best-known test drivers—who also were racing drivers—he lacks the fire of Sanesi but has the steadiness of Guidotti. And all three have that analytical, critical judgment that it takes to arrive at a really refined result. A car that does not hold the road to perfection goes right back to the shop. If there is any shortcoming in high-level performance, if there is any defect whatsoever, if the car is not wholly sincere, Ferrari is capable of

Boxer prototype number two.

feeling the fault, and he commands that it be eliminated. This is why it takes no time at all for a good driver to learn to handle a Ferrari, right out to the limits and without fear of ugly surprises. There are some fast cars that take two or three months to learn to control. And there are others that just leave the road and no one ever knows why. But a good driver is at home in a Ferrari from the word go. And it should be mentioned that that tradition and that mentality are still very much alive at Alfa.

"What makes Ferrari run?" Jano continued. "It's not the desire to make a lot of money. He is interested in having the cars that bear his name cut a beautiful figure in the world—*far bella figura nel mondo*. I think that he puts all of his money back into the business. If he would stop racing in Formula One he would gain 150 million lire per year, at the very least. But as long as he has the money to spend on racing he will do it. He never shirks in spending money on research and experimentation and his technical staff is large out of all proportion to the size of the business. He started with nothing. Then came the Scuderia, with garage space for a mere five or six cars. And look at what he has created. He has no other satisfaction. His family, his very life, is that creature of his, La Ferrari." ✠

ENZO FERRARI'S PERSONAL RACING RECORD		MILES	PLACE
1919	Targa Florio	268	9
1920	Parma—Poggio di Berceto hillclimb	33	3
	Targa Florio	268	2
1921	Targa Florio	268	5
	Circuit del Mugello	40	2
1922	Targa Florio	268	DNF
1923	Circuit del Savio	167	1
	Targa Florio	268	DNF
1924	Circuit di Pescara	158	1
	Circuit del Savio	167	1
	Gargnano-Tignale hillclimb	6	1
	Corsa sulle Torricelle hillclimb	3	2
1927	Circuit di Alessandria (Touring class)	99	1
	Circuit di Modena	224	1
1928	Circuit di Alessandria (Sports class)	99	1
	Circuit di Modena	224	1
	Circuit del Mugello	40	3
1930	Circuit di Alessandria	99	3
	Coppa Presolana hillclimb	5	3
1931	Circuit Tre Province	80	2
	Bobbio—Passo del Penice hillclimb	8	1

CHAPTER 2

A Certain Mystique

by David Owen

David Owen, Welshman, yachtsman, graduate engineer, editor, writer, contributed his first article to AUTOMOBILE Quarterly–*the story of Ferdinand Porsche and his cars–in 1970. He has been a contributor ever since. Mr. Owen brings a thorough background to his work, having been motoring correspondent for the* Sunday Times Magazine, *editor of the monthly magazine* Sporting Motorist, *scriptwriter for a weekly motoring program on BBC Television and contributor to the* Observer Magazine *and* Daily Mirror Magazine, *among others. A long-time admirer of Ferraris, Mr. Owen here describes some of his experiences with them and the rather special emotions these encounters have evoked.*

A hot, breathless afternoon in the countryside outside Bologna. The road was narrow, with a sheen of white dust, and in both directions from the right-angled bend where our car had stopped, it ran arrow-straight between the flat fields. The engine's death had been sudden and final, hinting at electrical trouble. We pushed the car off the road into the entrance to a farm track, and as we stood there wondering what to try next, a black-robed priest nodded his way past on a creaking bicycle. At the time, it seemed a much more sensible piece of transport than any modern, temperamental, unreliable, broken-down heap of automotive junk. And then we heard it: a distant, high-pitched insect whine, reaching over the far horizon. It swelled, first to a hum and then to a scream, growing closer and closer and louder and louder. On it came, trapped in a boiling shroud of dust. Just past us, the

priest had stopped. He climbed down off his bicycle with ponderous dignity, and wheeled it very deliberately into the waist-high grass at the edge of the road, before craning his neck to see what was coming. He *knew*.

It was closer now, near enough to see something under the dust cloud: a gleam of dull metal, a glint of sunshine on glass and chrome. The noise shivered at its peak and then died abruptly as the driver slowed for the bend. A rapid-fire volley from the exhausts and a bellow as he dropped into fourth, then into third, then into second, a puff of smoke from the wheels and then a long-drawn-out howl of sorely tried rubber as he slithered into the bend, swinging in close to clip the apex a yard from where we stood. A blur of silver and black, a glimpse of chamois-gloved hands on the polished wheel before the rising turbine scream of the engine blotted out everything else as he fed the power in again, balancing the car on the tiptoe edge of adhesion. Already it was fishtailing slightly as it vanished into the distance, and all that was left was a veil of dust hanging in the noonday sunshine, two crisp rubber marks on the tarmac, the priest shaking his head in wonder, and the memory of a badge, a prancing black horse on a yellow background. A Ferrari . . . . .

The next time was a misty November morning in the plains of Lombardy. It was the week of the Turin Motor Show, and pictures of the new cars had to be dropped at Milan airport to be put on the afternoon flight to London. My MG sat happily enough in the fast lane of the autostrada, in overdrive top gear with a hundred and five on the clock, long ranks of rust-brown poplars flicking past in the mist. It was still early, and traffic was light. Then, half a mile back, through the rear-view mirror, came a warning. An orange flash of headlamps, repeated once. I eased the wheel a fraction, and the MG edged obediently into the center lane and I heard, a split-second later, the jarring wail of twin windhorns and BLAAAAAAAAAARGH! A blood-red 250 GT Ferrari shrieked past, its bow-wave rocking the car as if we'd been standing still.

Third time lucky was only a day or two later. An offer of a lift from a fellow journalist between the Motor Show and the railway station was welcome enough. When it turned out he had bribed someone into letting him loose in a 330 GT Ferrari, the ride became top priority. Conditions were less than promising: traffic was heavy and the sidewalks crowded, but whatever the circumstances, no one with blood in his veins could drive one of these slowly. Almost without trying, the speedometer needle edged up to the ninety mark, even in narrow sidestreets. Entering intersections, we found windhorns were not needed—a quick blip on the throttle and the wonderful boom of a racebred V-12 trying to slip the leash made the shoals of commuter Fiats part in front of us like herrings before a shark.

Every junction had piles of wet leaves over a foundation of

streetcar tracks and greasy cobblestones, but the Ferrari took it all in its stride with never a falter, never a twitch, whatever the radius and whatever the camber of the surface. Cushioned in the leatherbound calm of the cockpit, the icy, wet, crowded, workaday world was remote as Jupiter. Everywhere faces turned to stare, with frank adoration or green-eyed envy. Even the police joined in, holding up crossbound traffic and urging us on with piercing whistleblasts and furious twirls of their white batons. Driving a Ferrari in Italy is a heady sensation: it feels a little like being Pope for a day.

This doesn't sound like a sober, unemotional assessment of a piece of automotive engineering, does it? I make no mention of brake pressures, power-to-weight ratios, speeds in gears, steering characteristics or weight distribution. All these things are relevant, as they are for every automobile ever designed: it's simply that the Ferrari isn't that kind of car. Inviting an enthusiast to look at a Ferrari objectively is like asking a man to comment on his beloved's IQ, or the state of her bank balance. At worst, it's impertinent; at best, it makes not the slightest difference to the way he feels about her.

Not everyone loves Ferraris. There is nothing starry-eyed about that body of experts which earns its living driving, testing, assessing and writing about cars. I've heard opinions expressed on some of the most famous names in the car world which should have sent their makers weeping all the way to the bankruptcy court. But ask them about Ferrari, and watch their reactions: sometimes affection, sometimes hatred, but always respect, and never indifference.

What is this car which has such an effect on those who know it? Even the name has a ring of drama in its sound, a scornful sneer of machismo, or the brooding bravado of the bullring. Legends grow everywhere around Ferraris—some say they are killers, pointing to the long and tragic list of top drivers who spent their last moments behind the wheel of a Ferrari. Others say the deaths were inevitable because Ferrari drivers entered more races for more seasons than any of their opponents. What no one can deny is that for twenty-five years, whether storming up the Raticosa Pass in the Mille Miglia, blasting down the Mulsanne Straight at Le Mans or high on the banking at Monza or Daytona, Ferrari has been a formidable name to conjure with. And above all, a car that nearly everyone, sometime in their lives, has fantasized about owning.

The first Ferraris I ever saw were Grand Prix cars. When I was growing up, weekends were for racing—riding or hitching a lift to circuits like Brands Hatch or Silverstone or Aintree, to watch drivers like Mike Hawthorn, Peter Collins, von Trips or Phil Hill race against sleek dark-green Vanwalls or the cheeky, bulbous Coopers. This was Formula One racing, the toughest echelon of a dedicated and professional sport, where victory means *everything*. And the Ferrari Dino GP cars were typical of the Ferrari breed—strong, fast, reliable.

But what makes Ferraris so impressive to me is that the same dedication, the same sophistication, the same single-minded pursuit of excellence regardless of cost which goes into Ferrari's racing cars, goes into every Ferrari which leaves the factory gates. While other manufacturers may have babbled at times about how their racing cars improve the breed as a whole, there has seldom if ever been the kind of fusion between race car and road car that one finds in a Ferrari. Which is perhaps why seeing a Ferrari driven in anger on the open road is a sight not easily forgotten.

What is the reality behind a Ferrari road car, behind its performance, beneath the almost invariably beautiful bodywork? The latest, the fastest and the most expensive of the breed—the $36,000 Berlinetta Boxer—is a good example. It squats close to the ground, its purposeful crouch betraying the brute power beneath the elegant tear-drop shape of the Pininfarina bodyshell. This is no ordinary Ferrari, if that's really the word I should be looking for—even by Modena standards, this is something else again. Its official description is a mid-engine, high-speed coupe, which is about as informative as saying that it has four wheels and runs on gasoline. ''Engine'' in this case means a mighty flat-twelve four-and-a-half liter, four-overhead-camshaft monster squatting ahead of the rear wheels, needing no fewer than six double-choke Weber carburetors to keep it adequately supplied with high-octane spirit. ''High speed'' means a knot or two short of the magic 200 mph, *even in road trim*. And ''coupe'' means a compact two-seat projectile with a 380 brake horsepower engine churning away at up to 130 spins a second within

inches of the driver's neck, all in the interests of balance and handling. All of this has to add up to something special.

And it does. As always, it isn't so much *what* Ferrari does—although this is always a great deal more than anyone else would dream of doing just to produce a motor car—but the *way* that he does it which makes all the difference. Setting up an engine to produce horses by the hundred is a fairly straightforward design exercise after all. But putting this kind of power on the road and keeping it there, requires engineering of the most sophisticated kind. Plenty of manufacturers already turn out over-powered stoplight GP specials with some of these characteristics. Where the real artistry comes in of course is in making such enormous performance part of a coherent whole—in setting up suspension, steering, chassis, brakes and weight distribution to cope with this magnitude of power and use it fully.

The result, for all of us who experience it for the first time, is shattering. Driving a car like the BB feels like driving any other well-designed, well-set-up modern sports car, except that the time scale is cut in half. The steering feels a little on the dead side at first until you remember it's designed for stability at three miles a minute rather than quick reflexes at parking speeds. But in the right conditions, a car like the Boxer is close to the furthest point the internal-combustion vehicle can reach from its cranky forebears of ninety years ago. A few months ago, I met two friends in Italy. They had driven from the Channel in one of these cars, a thousand miles across Europe through heavy traffic, rain and gathering darkness in a fraction under eight hours—an *average* of 126 mph! And they

stepped out of the car ready to tackle the return journey straight away, if only the owners could be persuaded to part with it for that long. With ever-stricter speed limits, ever-heavier traffic and ever-scarcer fuel, it's difficult to see how much further the superfast, super-luxury coupe can go along this particular track—so the Boxer could well end up as the ultimate Ferrari.

But in the end we all have our own personal images which the name Ferrari conjures up for us. For me it isn't the fastest, biggest or the most expensive model I remember with the most affection. Rather, the Ferrari for my money is the lovely little Dino coupe—just 2.4 liters in a sleek Pininfarina two-seat body, as carefully shaped as a piece of sculpture. Nothing very remarkable in size, in weight—or, to be brutally honest, in performance either—yet precisely because it *is* so comparable in cold figures to sports cars turned out by several other makers, I find it my favorite. Sit in it, drive it, and you have to admit that even a smaller, slower Ferrari like this has a liberal helping of the family character in its pedigree.

First of all, you notice the noise. You can produce 195 horsepower simply enough by doing what most manufacturers do—either by making your own 3.5 liter engine, or by buying a bigger, lazier unit from Ford or General Motors. Ferrari does things differently: using less than 2.5 liters in the Dino, he puts in a V-6 four cam with three twin-choke Webers, a jewel of an engine descended directly from the Championship-winning 1.5 liter Formula One power unit of 1961. And, though the output may be the same whichever way you choose, the sound is a complete giveaway, from the second you start up. There's no other noise on earth like that Ferrari signature tune: the deep, resonant boom from the big-bore pipes, the clatter of camshaft chains and the gnashing of valves. Anyone who could record it for playing back in Fords and Chevrolets would make a fortune . . . .

But the noise is merely an overture: you only sense the car's deeper character when it gets the bit between its teeth. At rest, little details like the sliding alloy gate for the gearshift lever feel awkward—first gear is set opposite reverse, with a move across and forward before you can hit second—and the clutch and brake pedals feel stiff and heavy. But once you build up speed, you can *feel* the car wake up and come alive—the steering lightens, and feels as direct and sensitive as if connected straight to your nerve endings, and the faster you snatch shifts from gear to gear, the easier and more reliable the change becomes.

Time was that Ferrari's highly tuned engines only gave of their best over a very narrow speed band—drivers said they had to be rowed along on the gearshift lever to get the best out of them. But today the Dino's superb little engine is as tractable as a Model T's—it will pull from only 1000 rpm in top if you're careful, and will spin up to almost eight times that speed with no sign of temperament or fuss,

beyond a steadily increasing howl of power which sets the blood tingling.

Some experts say that mid-engined cars are all very well on the track—because their low polar moment of inertia and even weight distribution gives them good roadholding and quick handling— but that their tendency to spin easily and a loss of grip that happens too quickly and completely on all four wheels makes them unsafe on the road. Yet hit a corner in the Dino and you go round as predictably and undramatically as if sitting in a streetcar. Take your foot off the throttle in the middle of a sharp bend and the car twitches slightly—drop into second and put your foot hard down, and the tail will break away. But in both cases the reaction comes predictably and consistently, and the steering is more than quick enough to allow you to correct in time. No traces of the old Ferrari surprises for unwary drivers.

There is one very good reason for this. Like the engine, the Dino's running gear is descended from the 1961 Grand Prix racer—coil springs and wishbones front and rear, with rack and pinion steering. Under its streamlined body, this is the nearest thing to a full-blooded Formula One racing car you'll ever be able to drive on a public road. Over the last twenty years, sports cars have lost a lot of the advantages they once enjoyed over production family models thanks to progressive design improvements. But Ferrari's eternal refusal to compromise restores the difference.

This is all very well if you are lucky enough to have a Ferrari of your own. But what of the rest of us, who can only envy from afar? Looked at from our point of view, *any* Ferrari is an extravagance, an expensive and utterly selfish toy for the pampered few who can afford it. Yet that doesn't alter the truth: that all of us, whether we drive a Ford or a Buick, a Beetle or a bike, feel this remarkable car's mystique and admire its performance. We *need* Ferrari. Too many cars today are designed by cost accountants—too many design advances never appear on production cars because they add a dollar or two to the selling price. This is inevitable in a highly competitive industry, part of the price we pay for reasonably cheap, personal transport.

Enzo Ferrari will never compromise to cut costs. Anyone as single-minded as he is about cars and racing will spare no expense to make his cars better. It's ruthless in a way and probably unbusinesslike. But business—like everything else—remains secondary to racing for Ferrari anyway. And it is probably good that it has. For Ferrari's obsession has produced some of the very best cars in all the world, cars against which others can be measured, cars that show just how much can be done when excellence is the only priority. In centuries to come, when the automobile is just a memory, our descendants could do worse than judge the achievements of the Motor Age by looking at the very best we could do. At a car like the Ferrari. ✤

49

CHAPTER 3

A Beginning -the 815

by Peter C. Coltrin

Seventeen years ago, Pete Coltrin stopped in Modena to see his good friend, the late automotive writer Hans Tanner. He has been there ever since. Pete Coltrin's first love was airplanes and he earned a private pilot's license in 1945 at age sixteen. It was while at college in California that he became enamored of automobiles and—under the watchful eye of Ak Miller who became a lasting friend—he prepared a 1949 Olds 88 for racing at Muroc Dry Lake, Bonneville and local drag strips. In 1957 he accompanied Miller and Miller's Chrysler-powered racing car to Italy for the Mille Miglia. A year later, Mr. Coltrin returned to Italy, married an effervescent Modenese woman, Gabriella, and became a contributor to several magazines including Sports Car Illustrated, Car and Driver, Road and Track, Motor, London, *and others. In this chapter, he takes an updated look at Ferrari's 815. The account is based on conversations with those involved in the project and with Enzo Ferrari, whose memory about the 815, Mr. Coltrin says, is a bit hazy now though he still fondly regards the car.*

The Vettura 815, a car conceived thirty-six years ago and realized and raced barely four months after conception, was, in fact if not in name, the first Ferrari. It was built to the design of *Ingegnere* Alberto Massimino by workmen at Ferrari's machine tool company, Auto Avio Construzione, on Viale Trento Trieste in Modena, premises which today house Ferrari's sales and service headquarters. Engineer Massimino along with Enrico Nardi and *Ingegnere* Vittorio Bellentani—both of whom assisted on the 815's development—had all joined Enzo Ferrari when he left Alfa Romeo

50

Lotario Rangoni and Enrico Nardi contesting the 1940 Mille Miglia in their

The Rangoni/Nardi 815 heads for

in 1938 and together formed the nucleus of Ferrari's new organization. Luigi Bazzi, who would play such a vital role in later Ferrari history, did not figure in the 815 project, for he remained at Alfa during the war working on aero engines.

Although Enzo Ferrari was commercially most concerned with machining airplane engine parts when he began his company, the visit of two young men to his factory one day soon led him back to racing cars. In late 1939, Alberto Ascari and a friend approached Ferrari and asked him to build a pair of sports racing cars for them. The twenty-one-year-old Ascari was the son of Ferrari's late race driver friend, Antonio Ascari, and with a successful motorcycle racing background behind him, he now wanted to try his hand at four wheels. Ascari's companion was the Marquis Lotario Rangoni Macchiaveli di Modena, a scion of one of the city's most distinguished families who had taken part in some local car races and whose garage was just up the street from Ferrari's premises.

The late Enrico Nardi remembered, when interviewed, that the decision to build the requested two cars was made on Christmas Eve, 1939, at a dinner party. And Alberto Massimino has perhaps best stated Ferrari's characteristic reason for embarking on the project this way: "Then, as now, his passion for racing cars had not abated nor was it on the wane."

The objective of the new car was to provide both Ascari and Rangoni with mounts for the upcoming Gran Premio di Brescia, an event scheduled to be run on April 28th, 1940. Generally referred to as the 1940 Mille Miglia, the race did not cover the Mille Miglia's traditional route around the upper mid-half of the Italian penninsula. In 1938 there had been some serious accidents during the event and the government had banned its running in 1939. Neither this nor the fact that much of Europe was already at war by the end of 1939, discouraged the Automobile Club of Brescia, however, and the original Mille Miglia organizers—chief among them Renzo

ng the 1940 Mille-Miglia. **Early trials of the 815.**

Castagnato and Conte Aymo Maggi—found a way around the government's edict. They organized a so-called "short Mille-Miglia" though it actually consisted of 933 miles of open roads, a triangle from Brescia to Cremona to Mantova to Brescia. There were five general classes: 750 cc, 1100 cc, 1500 cc, 2000 cc and 3000 cc sports racing cars.

Ferrari and his collaborators chose to contest the 1500 cc class for several reasons. As time was short, they knew that many already available parts would have to be used if their car were to be finished by April. And such parts would be readily adaptable to a 1.5 liter car. Then, too, Fiat was offering cash prizes for class winners using Fiat components. Within the limits of incorporating many pre-existing bits and pieces, Massimino displayed considerable ingenuity in the new design. He based much of the car on the Fiat 508 C Ballila. The Ballila in both coupe and sedan versions had not only proved popular with the motoring public but also with specialty builders of the day.

Many *spinta*—literal meaning, pushed—versions, including some with special bodies, had run in local races, hillclimbs and the Mille Miglia. Today we'd think of them as hotrods. The 508 C was a tuner's delight, perhaps akin to the robust Fords and Chevrolets that were being modified at the time in the United States. Rangoni had a 508 that had been tuned by Modena Fiat dealer and race car enthusiast Vittorio Stanguellini.

There were good technical reasons for basing a new racing car on the Fiat. As a production car, the 508 C was abreast if not ahead of its time. It had, for one thing, a 1089 cc overhead-valve four-cylinder engine and four-wheel hydraulic brakes at a time when some of its contemporaries were still using side valves and mechanical brakes. The front suspension was independent, DuBonnet type with coil springs and hydraulic shocks in an oil bath. Rear suspension was by a rigid live axle suspended by semi-elliptic leaf springs with hydraulic shocks and an anti-roll bar. Both front and rear suspensions were

53

On the road to Maranello, Ernesto Maserati (under sign) watches 815 test runs in February, 1940.

well-placed and rigid, and the car was fitted with Borrani wire wheels.

The main problem Massimino faced was the engine. The Fiat's 68 X 75 mm bore/stroke dimensions were impossible to stretch out to one and-a-half liters. Thus Ferrari and Massimino decided to create and build their own straight-eight engine. Again using as many Fiat components as possible, Massimino designed a one-piece block with iron wet liners and a ribbed aluminum sump. The block, sump and valve covers were cast by Fonderia Calzoni in Bologna, Ferrari not yet having a foundry of his own. But Ferrari's shop did make the sixteen lobe camshaft and the five-main-bearing crankshaft. Valve rockers, shafts, and connecting rods were Fiat components. Atop the engine's block, Massimino placed, in line, two 508 C cylinder heads. He replaced the Fiat's pair of hard-to-synchronize distributors with a single Marelli unit which also drove the tachometer. A new water pump was made to handle the engine's increased capacity over the Fiat.

The new little eight was aspirated by four Weber 30 DR 2 downdraft carburetors mounted between the exhaust manifolds which led to a single exhaust pipe. Bore and stroke were 63 X 60 mm, and compression ratio was raised from the Fiat's 6:1 to 7:1. With its 1496.28 cc displacement, the engine developed some 72 to 75 hp at 5500 rpm. It would be enough to propel the car to 112 mph. This engine gave the new race car its name for, in lieu of putting his own name on the car—something his contract with Alfa had forbidden for a period of years—Ferrari instead designated the machine with a blue and yellow badge that read 815, for eight cylinders, 1.5 liters.

Two bodies were built for the chassis by Carrozzeria Touring in Milan and, like some of Touring's other creations from the late Thirties, these hinted at the envelope body with its integrated fenders which would become universal after the war. The rather stylish coachwork was relatively smooth and, though pleasant-looking, it was not unique. The 815's appearance was very similar to Touring's Alfa-Romeo spyders. Both 815's—numbered 020 for Rangoni and 021 for Ascari—were completed in enough time to be tested on the autostrada and on some Emilian side roads. Enrico Nardi was chief test driver, this being his primary role in the 815 project. "I only tested them," he told me in 1965. "Alberto Massimino designed

Albert Ascari and his friend, Sig. Spoldi.

During World War Two, Ferrari produced the brochure (left) describing the products of his company, Auto-Avio Costruzioni. Though racing cars were not among the firm's products, the pamphlet's cover depicted two 815's and the emblem Ferrari would use on the first car to bear his own name. Below: the 815 up close.

The shortened 1940 Mille Miglia was held on a roughly triangular course in the Brescia area and was also known as the Brescia Grand Prix. A map of the course together with a listing of past Mille Miglia winners was posted at a Brescia restaurant where it drew much attention.

815

VETTURA "815" SPECIFICATIONS

Constructor • Auto Avio Costruzione • Modena (Ferrari)
Engine • eight-cylinder • in-line • one-piece cast aluminum block

Bore x Stroke	63 mm x 60 mm
Displacement	1496.28 cc
BHP (CV) at RPM	72-75 at 5500 RPM
Cylinder heads	2 aluminum Fiat Tipo 508 C ohv.
Crankcase	wet sump, one piece alloy casting
Crankshaft	steel—2-4-2 configuration—five main bearings (Fiat anti-friction)
Valve actuation	pushrods and rocker arms
Carburetion	4 Weber Tipo 30 DR 2 single-throat down draft
Lubrication	force feed, two pumps in sump
Clutch type	single dry plate
Gearbox	four speed and reverse—first and second non-synchronized, third and fourth synchronized, "H" pattern
Driveline	Cardan driveshaft
Brake type	Fiat hydraulic drums, front and rear
Wheels	3.25-15 Borrani center-lock wire
Tires	Pirelli "Stelvia" 5.50-15
Steering type	Fiat worm and sector (508 C)
Top speed	112 mph
Chassis	Fiat 508 C Balilla 1100 reinforced channel section cruciform
Suspension	Front: independent parallelogram, coil springs and shock absorbers in oil-filled casings (Dubonnet system)
	Rear: live axle; semi-elliptic leaf springs, hydraulic shock absorbers and transverse stabilizing bar
Dry weight	1217 lbs.
Wheelbase	95.2 inches
Track, front and rear	48.4 inches front, 48.2 inches rear

them.''

Although the cars were privately owned, they formed, in effect, a Ferrari team. Today memories are rather vague about the team's pre-race strategy. Some recall that there was an agreement to take it relatively easy on the first few laps, especially since the 815's soon proved themselves to be clearly the fastest cars in their class. Perhaps passion won over discipline.

At the end of the first 110-mile lap, Ascari led the 1500 cc class, followed by Rangoni. Then, on lap two, Ascari retired with a burnt or broken exhaust valve, possibly caused by rocker arm failure. It was an unexpected source of trouble because these Fiat components had always proved most reliable. After Ascari's retirement, Rangoni carried on—with Nardi as his co-driver—at the head of the 1500 cc cars. He was leading by thirty-three minutes when he too broke down. A contemporary report said that a small bearing had failed. Nardi claimed it was a timing chain, at least to the best of his memory.

In any case, the race was over for the 815's, and very soon all racing was finished as war engulfed Europe. During the conflict, both the Ascari and Rangoni cars were carefully stored away out of sight and almost out of mind. When Rangoni was killed testing a three-engine bomber, his car became the possession of his brother. Unfortunately, it was broken up for scrap in 1958, a fate about which Rolando Rangoni is most regretful.

It was believed for a time that Ascari's 815 had met a similar end, and it very nearly did. The car was acquired by a Milanese collector named Belatracchini who raced it at Pescara on August 15th, 1947. But Belatracchini did not complete the race, received little notice in contemporary reports of the event, and sold the car soon afterwards. Eventually the 815 somehow found its way to first one wrecking yard and then another until it was rescued—on the day it was scheduled to be broken up—by Signore Emilio Fermi-Storchi, curator of a small private museum in a village near Modena called San Martino in Rio.

Fermi purchased the car for $150.00 and spent a great deal more on restoring the machine. A neighbor of Fermi's, Giacomo Trullo, was an expert at resurrecting engines and bodywork and brought 021 back to as nearly new as possible. While Trullo has had to take a few liberties with the car's headlamps—using surrounds from a 1939/40 Opel, the car's flavor and character remain unchanged.

Ferrari himself has written that the 815 was a failure because of the necessarily short development time it received. And yet, after these many years, I know that he is happy that one of the cars still exists. Personally, I think he is being too hard on the 815. In a way, it pointed to the future, having led the race for which it was designed. I think Ferrari and his engineers did a fine job with their first car. But then perhaps, I'm prejudiced.

The Red Cars
-the early years

The first decade of Ferrari cars represents one of the most intriguing periods in the marque's history. More than half a dozen different coachbuilders were then at work clothing Ferrari chassis in a svelte array of body styles intended for racing or touring. In doing so, they crafted some truly forceful and enduring designs. Readers will doubtless have their personal favorites among the cars presented on the next twenty-four pages but it is safe to say that a trio of these cars deserves special attention for managing to fulfill with complete success the demands of their genres—the open sports car and the sports racer. First there is the lithe and tiny, sweetly curved 166 Mille Miglia roadster by Touring, a car which remains fascinating and fresh a quarter century after it first took the road. Next is the 410 Sport by Scaglietti, its paper-thin aluminum sculpted into aggressive scoops and bulges which leave one unsurprised at the shrieks loosed from its four big-bore exhaust pipes. Finally, there is the low and graceful Testa Rossa, as lovely a racing car as can be imagined, a classic of its type. The best way to see these cars, of course, is up close where they can be touched, smelled, and listened to. In lieu of that great pleasure, we hope that this portfolio (which represents the largest collection of Ferrari car portraits ever published) will be a fulfilling and thought-provoking substitute. (Ed.)

1940 Auto Avio Costruzione 815 with coachwork by Touring ● Owner: St. Martino in Rio Museum

1947 166 Inter ● Owner: Henry Austin Clark Jr.

1950 212 Mille Miglia with coachwork by Touring ● Owner: Giulio Dubbini

1950 166 Mille Miglia with coachwork by Touring • Owner: Larry Taylor

1950 195 with coachwork by Vignale ● Owner: John R. Quee

51 212 Export with coachwork by Vignale ● Owner: Tom Oleson

1951 212 Mille Miglia with coachwork by Vignale ● Owner: Johnny Johnson

1951 212 Export with coachwork by Vignale • Owner: Tom Oleson

225 Sport with coachwork by Vignale ● Owner: David Shute

1953 250 Mille Miglia with coachwork by Pininfarina ● Owner: Kirk F. White

1953 375 America with coachwork by Vignale ● Owner: Wayne Golomb

1954 212 Inter with coachwork by Ghia ● Owner: Tom C. And

54. 860 Monza with coachwork by Scaglietti ● Owner: David Piper

1954 Mondial Series I with coachwork by Pininfarina ● Owner: Robert N. Dusek

1955 410 Sport with coachwork by Scaglietti • Owner: Robert N. Du

1957 Testa Rossa 500MD TRC with coachwork by Scaglietti • Owner: Helmut T. Brandt

1956 250 GT Tour de France with coachwork by Scaglietti ● Owner: Robert N. Dusek

1957 250 GT with coachwork by Boano • Owner: Jim Trott

1958 250 GT Tour de France with coachwork by Scaglietti • Owner: John Gaughan

1958 412 MI with coachwork by Scaglietti • Owner: Steve Earle

Photo
Credits
for this
Section

61, 63: Photographs by Giorgio Boschetti. 62: Photograph by Henry Austin Clark, Jr. 64, 65, 76-77, 78, 79, 80-81, 82: Photographs by Stan Grayson. 66, 67, 68, 69, 73, 74, 83 bottom, 84: Photographs by Marc Madow. 70, 71, 75: Photographs by Neill Bruce, Kingsclere, Newbury RG15 8TD, England. 72: Photograph by Don Vorderman. 83 top: Photograph by Rick Lenz.

The First Ferraris

by Stan Nowak

Franco Cortese debuts the Ferrari 125 at Piacenza; May 11th, 1947.

Stanley Nowak is a New Yorker with a lifelong love for cars. A graduate of the Art Center School in Los Angeles, Mr. Nowak has worked in and around the automobile business for years. He is a member of the Society of Automotive Historians and his articles have been published in AUTOMOBILE Quarterly as well as other magazines. Mr. Nowak has restored more than twenty historic cars, and his book, Automobile Restoration Guide, *is now in its second printing. A long-time interest in Ferraris led to his purchase of the first two cycle-fendered cars and he supervised their restoration. One of those cars–today owned by an English collector–is the oldest Ferrari in the world. Among the other Ferraris Mr. Nowak has owned was the Tipo 212 once raced by Phil Hill. Drawing on his experience with the earliest of Ferrari automobiles, Mr. Nowak here describes in detail how these cars came to be and what they were like to drive.*

The fastest way to Modena is by jet to Milano and then by rented car south on the autostrada. In twenty minutes you pass the exit for Piacenza, the city that marks the northern edge of the region of Emilia and the site of the very first public appearance of a Ferrari automobile in May, 1947.

The autostrada slices southward through the heart of Emilia, broad flat country noted for the richness of its farmland and for the great number and variety of its small industrial enterprises. In the heart of Emilia is Modena, less than two hours drive from the airport in Milano (although the trip has been made in a Daytona in forty-seven

minutes!). Here the great marques of Maserati and Ferrari have flourished. Undoubtedly, the success of these cars has been greatly assisted by the nature of the men of Emilia, noted for their volatile, aggressive approach to their work. Enzo Ferrari expressed very well this dubious asset in his autobiography: ''The worker in these parts, whether of muscle or of mind, is extremely intelligent and very active. In short, the union of blood and brain is such that the result is a type of man who is stubborn, capable and daring—the very qualities that are needed for making racing cars.''

When one arrives in Modena he will probably stay at the Real Fini or the Palace. Either of these fine hotels, long popular with *Ferraristi* from throughout the world, is within a block of the Ferrari factory sales office which also serves as the delivery and service center for foreign visitors. This 'downtown' facility is, in fact, on precisely the same site where the original Scuderia Ferrari designed the 1.5-liter, straight-eight supercharged, single-seater racing car that was to become world famous ten years later as the unbeatable Tipo 158 Alfa Romeo.

Except for the superchargers and the engine liners, the Tipo 158 racing cars were built entirely by Ferrari's team in Modena. Enough parts were produced to construct at least nine engines and six complete cars, and the bodies were made by Carrozzeria Zagato as required. Thus, Enzo Ferrari enjoyed his first experience as the manufacturer of a new line of racing automobiles. Scuderia Ferrari had, of course,

previously been responsible for the development of new models but these were always revisions of existing cars. For example, the 8C 2300 supercharged Alfa Romeo sports car on the short chassis was transformed by Ferrari into the much faster Monza model by shortening the wheelbase, enlarging the engine and fitting ultra-lightweight coachwork.

By the end of World War II, Ferrari had moved his small factory to the town of Maranello, just outside of Modena. The move was forced on him by an Italian law requiring the decentralization of new industries and, as he owned a parcel of land in Maranello anyhow, he decided to use it in this manner. It was here, in 1945, that he began to lay plans for the first automobile that would bear his name.

Essential to his success was a brilliant engineer, and Ferrari chose Gioacchino Colombo, the same man who had created the Tipo 158 Alfa Romeo for him in 1937. In addition, he would rely on his old friend Luigi Bazzi to develop and perfect whatever Colombo designed. The idea of a 1.5-liter V-12 engine was daring in the extreme. In the war-torn Italy of 1946, still suffering from extreme shortages and erratic supplies of essential products, Ferrari was warned that the ambitious V-12 engine project would be a failure and would ruin him financially. While there has been much speculation on this subject, there can be no doubt that the choice of the V-12 layout was solely Enzo Ferrari's. But it was then up to Colombo to produce a rational

On this page, three 125's.

design that could enable Ferrari to keep the initial tooling and production costs at a reasonable level.

Colombo's experience with multi-cylinder engines dated back to the V-12 Alfa Romeo 12C of 1936, a single-seater designed by Engineer Vittorio Jano, with Colombo's assistance. This particular design developed 370 hp at 5800 rpm from a supercharged engine of just over four liters in size. One of the 12C cars had the distinction of winning outright the Vanderbilt Cup Race of 1936 driven by Tazio Nuvolari.

While still working for Scuderia Ferrari in Modena, Colombo also designed and built the fastest prewar Alfa Romeo Grand Prix car, the supercharged Tipo 316. Its engine appeared to be a V-16 but in reality it was a W-16 as each bank of eight cylinders had its own crankshaft. These were geared together in a common crankcase. The top sections of the engine were actually Tipo 158 parts. The bore and stroke were 58 mm x 70 mm (the same as the Tipo 158), displacement just under three liters, and the engine developed 440 hp at 7500 rpm. The Tipo 316 was Alfa's last attempt to match the power and performance of the German Auto Unions and Mercedes-Benz. Its first appearance was at the Tripoli Grand Prix of 1938 where it established the fastest lap during time trials at 136.5 miles per hour. It did not run in the race, which was won by Hermann Lang's Mercedes at 127.45 miles per hour. Later in the year, however, Farina finished second overall in the same Alfa in the Italian Grand Prix at Monza, defeated only by Tazio Nuvolari driving an Auto Union.

Certainly, Colombo's technical background and his prior experience with the Alfa Romeos and with Ferrari himself made him the ideal man to design a new V-12. Ferrari's first announcement of his new cars came in late 1946. The headline at the beginning of the article in *Inter Auto* read ''La Nuova '125 Sport' Della Scuderia Ferrari.'' It gave full specifications for a vehicle called the ''Ferrari 125 S'' but advised that two further versions, *Competizione* and *Gran Premio* would be announced shortly. Eventually, Ferrari issued a small folder which gave specifications for the three models. All of them specified a V-12 engine with a bore and stroke of 55 mm x 52.5 mm totaling 1496.77 cc. Each cylinder had a cubic capacity of approximately 125 cc and this figure became the type number. With few exceptions this system has been used by Ferrari for all his models since.

Actually none of the announced models were ever produced with exactly the specifications listed in the folder but the categories did indicate Ferrari's philosophy of producing three types of cars: single-seater racing cars for the Grand Prix formulas, two-seater sports cars designed to win races within particular classes and, lastly, high speed grand touring cars for wealthy customers with no interest in formalized racing but with a strong desire to bask in the reflected glory of Ferrari's racing successes. This approach continues to the present time.

Unquestionably, the new Ferrari evolved directly from its Alfa Romeo ancestry. Had circumstances between Enzo Ferrari and the management of Alfa Romeo been untroubled, the new V-12 might well have been another Alfa Romeo! Surely, this would have been the case had Alfa not hired the Spanish engineer, Wilfredo Ricart, who was the cause of the split between Alfa Romeo's General Manager and Enzo Ferrari.

In the rebuilt factory in Maranello—it had been bombed twice during World War II—Colombo's design was being translated into metal, and parts enough for at least six cars began to arrive from outside suppliers in mid-1946. Ferrari did not yet have his own foundry and the aluminum cylinder block was cast—as the 815's had been—by Alessandro Calzoni in Bologna. The oval tube X-frame chassis—a proven feature of the Tipo 158 Alfa Romeo—was built by Gilberto Colombo (no relation to Engineer Colombo) of Gilco Autotelai S.R.L. in Milano.

By the end of 1946 two Tipo 125 engines were running on the dynamometer and two chassis were being assembled to accept them. The engine development work was by now primarily in the hands of Aurelio Lampredi and Ferrari's longtime follower, the remarkable Luigi Bazzi who had been with Ferrari since 1926 when he left Fiat for Alfa at Ferrari's urging. Bazzi's genius lay in his ability to

1948: Tazio Nuvolari at the wheel of a 166.

improve and perfect and, if necessary, redesign in order to eliminate problems. He had proved himself at Alfa Romeo, Scuderia Ferrari and Alfa Corse, always under Ferrari's management. His contribution was almost always unsung and unheralded, and yet his work was perhaps equal in importance to that of the original designer.

Throughout the winter of 1946 and into the spring of 1947 Ferrari's engineers struggled to prepare two cars for 1947's racing season.

Technically, the 125 S shaped up as follows. At the front, the suspension was independent by unequal length A-arms with an elliptical spring running transversly across from one bottom A-arm to the other. The rear suspension utilized semi-elliptical springs on each side and an antiroll bar ran across the frame through one of the frame tubes. The shocks, front and rear, were unusual—similar to the 815—combined with a coil spring operating enclosed in oil. Later the rear shocks were sometimes removed and replaced by prewar style friction shocks. Steering was by worm and peg.

The first cars were built to Tipo 125 *Competizione* specifications, which meant using three Weber 32 DCF carburetors and ignition derived from two large Marelli magnetos driven off the rear of the camshafts. These magnetos were mounted horizontally and extended through holes in the firewall into the cockpit! By mid-1948 the magnetos were mounted vertically and driven by bevel gears. A five-speed gearbox was used in which fifth gear was an overdrive ratio. The brakes were large finned drums of cast aluminum with steel liners. Two shoes, one leading and one trailing, were used which resulted both in very heavy pedal pressure and lack of "feel." The wheelbase was 98 inches. The front tread was 50 inches and the rear tread 48 inches. Other than some engine parts and the frame, all major components were designed and built under Ferrari's supervision. The only major outside suppliers were Marelli for all electrical items and Pirelli for tires. To this day all major items in a Ferrari chassis not made at the works are made exclusively to Ferrari's design. Today Ferrari is a success and can afford this luxury, but in 1946 he might have avoided the time and expense by using standard components from outside specialists such as gearboxes, rear axles and steering boxes. But then, perhaps, his cars would not have been so special.

One can only speculate about details of how Enzo Ferrari was able to finance this project for such a long period of time with no apparent income. As in any major racing operation, there is always a curtain of secrecy drawn over the details of sources of capital. According to Luigi Chinetti though, Ferrari's wife Laura provided both capital and considerable know-how. Certainly, Ferrari had also amassed some capital of his own from his successful wartime machine tool business but, undoubtedly, he also drew upon the resources of his contacts, both private and commercial, throughout Italy. This may have

The first Ferrari ever sold, Besana's 166.

1949: Bonetto, Ferrari, Varzi, Bignami, Fangio.

Pescara, 1948.

included private grants from wealthy enthusiasts, extended lines of credit and other special terms from suppliers, as well as technical and financial assistance of some kind from Italian Esso and later Shell gasoline, Fiat, Pirelli and Englebert tires, Weber carburetors and others. Another source of funds could have been generated through the enthusiasm of certain talented race drivers who were also wealthy gentlemen most anxious to drive Ferrari's latest creation.

When the first two Ferrari chassis were completed in the spring of 1947, one was sent up to Carrozzeria Touring in Milano where it received a full-width roadster body of aerodynamic silhouette. This body design set no future trend but it was modern for its time and was bound to attract attention. Most importantly, on its nose was the now famous Ferrari badge, this one with the prancing horse's hoof touching the bar made by the extended top of the letter "F." Only a few cars later, the horse was moved slightly above the bar and it has appeared this way ever since.

The second chassis was fitted with a simple, rather boxy two-seater body—probably made locally—and cycle fenders. A few months later a third chassis went to Carrozzeria Touring, and this received full-width roadster coachwork almost identical to the first car's.

In early May of 1947, Ferrari entered the first Tipo 125 *Competizione* full-bodied car in a sports car race at Piacenza with the great champion Giuseppe Farina as driver. On the day before the race, however, Farina damaged the car in a road accident. An all-out effort was made by Bazzi and the Ferrari mechanics to repair and test the car in time for the race but, after all their work, Farina claimed himself to be unable to drive "for personal reasons." Whatever the reasons, Farina did not drive for Ferrari again for the rest of the 1947 season.

Franco Cortese was now called upon to race the new car in its first public appearance. Cortese was one of the old guard of drivers who had specialized in racing sports cars in Italian events and had been associated with Scuderia Ferrari before the war. During the first part of the thirty-lap, sixty-mile Piacenza race, Cortese stayed in the middle of the pack unable to reach the 6800 rpm required to develop the engine's full 118 hp. It was later discovered during a pit stop that this lack of rpm was caused by an overfilled sump. By the halfway mark, Cortese began to increase his speed and soon was gaining four to five seconds per lap on the leaders. A few laps later, he was in their slipstream and about to pass into the lead when the car's special aircraft-type centrifugal fuel pump seized. It was a stupid failure but the car had impressed Italian automotive experts with its strong V-12 engine. Its indifferent road holding was not mentioned. The race was won by a Fiat 1100 Special driven by Nando Righetti who was later to join the ranks of first-season Ferrari drivers along with Cortese, Tazio Nuvolari and Raymond Sommer.

93

Two weeks later Cortese took the cycle-fendered 125 to Rome's Caracalla circuit and finished first overall against a large but uncompetitive field. Throughout June, in either of the two available cars, Cortese competed in five different events and won first overall in three of them at Vercelli, Vigevano and Varese. The car failed to finish at a second Caracalla race and in the Mille Miglia. In July 1947, Tazio Nuvolari joined the Ferrari team. Perhaps it should be said that Nuvolari rejoined Ferrari, for he had been the number one driver in Scuderia Ferrari's Alfas for many years prior to the war. Nuvolari's first race in a Ferrari automobile was a small event at Forli, the Grand Prix of Arcangeli. There the little cycle-fendered Ferrari managed to withstand the rigors of Nuvolari's energetic driving technique and he ran away with an overall win. The following week, Ferrari entered both Nuvolari and Cortese for a larger, more competitive meet at Parma, only forty miles from Modena. It was, indeed, a memorable weekend, for Nuvolari's declining health was still fairly good and he was an enthusiastic participant in the festive activities which had started the evening before at the Winter Garden in the Ducal Park on the occasion of the ''Festa del Sorriso.'' A beauty contest was sponsored by the local Press Association and Nuvolari delighted in participating as one of the judges who selected Miss Parma and Miss Press Association.

The festivities were not over after the beauty contest, of course. The following day Nuvolari and Cortese in that order defeated a strong field which included Guido Barbieri in a cycle-fendered Maserati A6GCS and Felice Bonetto in a Cisitalia. Nuvolari led from the start and the ecstatic crowd broke through the barriers to congratulate the Flying Mantuan. Thinking quickly, Nuvolari stopped only long enough to pick up Miss Parma and disappear in a cloud of Pirelli tire dust! That evening, the great Tazio appeared again to take part in the Victory Banquet with Miss Parma. The Winter Garden rang with cheers for the winners and *Il Maestro*, as Nuvolari was referred to in the Italian press, had overnight established Ferrari as the new standard for sports/racing cars in Italy. It was quite possibly Nuvolari's last completely happy weekend of racing as his health began to deteriorate badly and, by the spring of 1950, it was impossible for him to drive at all.

A week later at Florence two Tipo 125 *Competiziones* were entered, but Cortese's broke down while Nando Righetti went on to finish third overall. It was obviously time to take stock as the Ferraris had been outclassed by larger cars such as Felice Bonetto's three-liter Delage. In addition, the race organizers were favoring a policy of dividing the classes at 1100 cc and 2000 cc rather than at 1500 cc. Further, a new Formula Two had been announced which called for unsupercharged cars up to two liters.

Barcelona, 195

i's winning 4.5 at Numancia corner during the Peña Rhin GP.

All this led to a decision at Maranello to increase the size and power of the 125 S engine. Thus, towards the end of July, 1947, Engineers Bazzi and Busso set about rebuilding two of the 125 *Competizione* engines. New liners and pistons were installed, giving a bore of 59 mm and new crankshafts were made, giving a stroke of 58 mm to achieve a volume of about 159 cc per cylinder and a total volume of 1901.87 cc. The enlarged engine was designated as a Tipo 159 and was installed in one of the old envelope-bodied cars. In its first appearance on August 15th, 1947, this machine—with Cortese at the wheel—finished second overall at Pescara. Lurking at the back of the grid was the ill-fated 815 with its new owner, Beltracchini, driving.

The Tipo 159 appeared again at Modena on September 28th where Cortese established the fastest time of the day but failed to finish because of fouled spark plugs. The winner that day was Alberto Ascari in one of the new A6GCS two-liter, cycle-fendered Maseratis. His mentor, the vastly-experienced Gigi Villoresi, was second in an identical car.

At this time, a second Tipo 159 Ferrari was being prepared which was probably based on the cycle-fendered car that Nuvolari had driven. In any case, the car was a new cycle-fendered, two-seater Spyder Corsa having a lower, more cigar-like silhouette and a more studied appearance overall than the 125. It is believed that this body was the work of Carrozzeria Touring. The car made its debut at the Turin Grand Prix for sports cars on October 12th, 1947, where it finished first overall driven by Raymond Sommer. It was a hopeful, auspicious conclusion to a somewhat uneven first season.

During the fall and winter of 1947-48 the Tipo 159 engines were enlarged again. This time the bore was increased to 60 mm and the stroke remained the same at 58 mm, giving an individual cylinder capacity of 166 cc and a total volume of 1992 cc. Thus, the Tipo 166 Ferrari was born, an event which marked the end of the unsupercharged 125.

The cycle-fendered Turin Grand Prix car became the first Tipo 166 Spyder Corsa and was sold to a wealthy enthusiast, Gabriele Besana, in January of 1948. It was Chassis Number 002C, Engine Number 002C and was the first Ferrari ever sold. It is probable that this car was a remanufactured version of one of the original full-with bodied Tipo 125's.

A second identical Tipo 166 Spyder Corsa was also built and delivered to Count Soave Besana—Gabriele's brother—in March, 1948. This was the second Ferrari sold and bore Chassis Number 004C, Engine Number 004C. Today, it is the oldest surviving original Ferrari.

The Tipo 166 Spyder Corsa continued to be built at Maranello

166 Spyder Corsa

during 1948 when at least six more of these cycle-fendered cars were constructed. They competed with success in both Formula Two—with the fenders removed and running on a gasoline/alcohol blend of fuel—and, with fenders, in sports car races. One of the new cars was built with full-width roadster coachwork by Carrozzeria Allemano and this car, driven by Clemente Biondetti and Count Igor Troubetskoy, won the 1948 Targa Florio, This same Allemano roadster was then rebuilt as a closed coupe and took first place overall in the 1948 Mille Miglia driven again by Biondetti, with mechanic Navone as his passenger.

At this point, the first Gran Turismo Ferraris were being completed, one of them going to Count Bruno Sterzi. It was a notch-back coupe by Carrozzeria Touring. Also among the earliest touring Ferraris were two designs executed by *Stabilementi Farina*. These were a convertible and a fastback coupe, both looking remarkably like their Cisitalia counterparts but slightly larger.

Nineteen forty-eight was a year of increasing activity at Ferrari as work was proceeding on development of the long-awaited supercharged Formula One car. In addition, concerted efforts were being made to win the most important Formula Two races. The cycle-fendered Spyder Corsas continued to see duty as both sports/racing and Formula Two cars. In May, Chico Landi, driving one of the Besana cars, recorded Ferrari's first Formula Two win at Bari against strong opposition from the little monoposto Cisitalias and Ascari and Villoresi in cycle-fendered A6GCS two-liter Maseratis, another dual-purpose Formula Two car of the period.

In July, Raymond Sommer won the Formula Two race at Rheims, held in conjunction with the Grand Prix of France, in a factory-prepared Spyder Corsa two-seater with Righetti in second place in a similar car. On September 9th, Count Besana missed a first place in the Formula Two race at Posillipo (near Naples) when he was outdriven by Villoresi in the incredible new 1100 cc OSCA.

It was on September 5th, 1948 that the first Formula One Grand Prix monoposto Ferraris appeared at the Turin Grand Prix. Three cars were entered but only Sommer finished—in third place behind Wimille in a Tipo 159 Alfa Romeo and Villoresi in a Maserati 4CLT/48. The Ferrari's performance was not encouraging. Initially the 1.5-liter V-12 supercharged engine was simply no match for the more

Ascari's four-cam 1.5 at Monte Carlo, 1950 is shown in the two photos below.

fully-developed Alfa.

At the end of September, Ferrari entered Sommer in a Formula Two race at Florence with a new single-seater Formula Two car. *Motor Italia* reported that the engine was developing 150 hp at 7000 rpm. No doubt, this was one of the Tipo 125 Formula One chassis with a two-liter Formula Two engine fitted. This was not difficult to achieve as the engines used identical blocks with the same motor mounts. It was, in fact, quite possible to race as a Formula Two car one weekend and as a Formula One the following weekend. This only serves to point up Ferrari's genius in utilizing the same basic engine for all categories of competition, something he has continued to do over the years whenever possible.

In the fall of 1948, Ferrari issued a brochure offering four types of cars for sale:

166 Sport — 90 hp at 5600 rpm, 1 32 DCF Weber carburetor.
    Illustrated by a notch-back touring coupe.

166 Inter — 110 hp at 6000 rpm, 3 32 DCF Weber carburetors.
    Illustrated by the Allemano-bodied Mille Miglia winning car.

166 Mille Miglia — 140 hp at 6600 rpm, 3 32 DCF Weber carburetors. Illustrated by a drawing of a yet-to-be-built touring-bodied barchetta roadster.

166 Formula II — 155 hp at 7000 rpm, three carburetors with type and make unspecified. Illustrated by a drawing of a single-seater Formula Two car.

The prototype 166 Mille Miglia— Chassis Number 0002M, Engine Number 022I—with its barchetta roadster coachwork by Carrozzeria Touring, was built in late 1948 and was shown at the Turin Auto Show complete with white-wall tires. It became possibly the only Ferrari ever delivered with such tires when it was eventually sold to the eccentric Los Angeles millionaire, Tommy Lee. This little car's wheelbase, incidentally, was 88.6 inches whereas the previous cycle-fendered Spyder Corsas had a 95.3 inch wheelbase.

The 1948 season had been one of substantial growth for Ferrari and was brought to most successful conclusion by the overall win of Luigi Chinetti in the Paris 12 Hours race for sports cars at Montlhery. His winning car was a Tipo 166 Spyder Corsa, Chassis Number 016I,

Pagnibon in his 212, winner of the 1951 Tour de France.

Engine Number 016I, and it was sold to Briggs Cunningham. It arrived in New York in June, 1949 and became the first Ferrari brought into the United States. Tommy Lee's barchetta arrived about a month later in Los Angeles.

The Tipo 166 Mille Miglia Ferrari was intrinsically a reliable machine. The engine, gearbox and rear end were virtually indestructible, and failures were almost always because of faulty components not manufactured by Ferrari or faulty assembly of the engine. The latter usually was caused by imperfect mating of the cylinder heads to the block due to the method of seating which involved O-rings around the cylinder bores and a peripheral gasket on the outer edge, rather than a conventional gasket. But properly set up, the car was not only surprisingly reliable, it was not expensive to race either. In Europe, these early Ferraris were frequently driven to a race over the road, raced and driven home.

The roadholding, ride and braking of the early Ferraris was less pleasant, however. Handling is characterized by oversteer which, if extreme, is at least predictable. The steering is heavy—even trucklike— and most of the early cars suffer from an almost incurable steering judder at low speeds—25 to 30 mph—which only a steering arm shock absorber can help to minimize. Brakes on these early cars seem nonexistent and pedal pressure is very high. Though the brakes improve when warm, they offer little reassurance. Ferrari did not lead the way in these areas, and he has acknowledged as much in his autobiography and in interviews over the years. Ferrari's great trump—his so called ''unfair advantage''—was and still is his engines.

The year 1949 saw the reemergence of Scuderia Ferrari as a full-scale racing organization. Two of the finest drivers in Italy were engaged to drive Ferrari exclusively: Alberto Ascari and Gigi Villoresi. The following list contains nineteen of the most important outright victories for Ferrari during 1949. Counting the smaller races and hill climbs, the total number of outright wins was thirty-two! Ferrari also gained nineteen seconds and twelve thirds!

Event	Car Type	Driver
Grand Prix of Rosario	Formula One	Farina
Targa Florio Tour of Sicily	Sports	Biondetti
Mille Miglia	Sports	Biondetti
Grand Prix of Brussels	Formula Two	Villoresi
Grand Prix of Luxembourg	Formula Two	Villoresi
Inter-Europa Cup	Grand Touring	Sterzi
Grand Prix of Rome	Formula Two	Villoresi
Grand Prix of Bari	Formula Two	Ascari
Grand Prix of Naples	Formula Two	Vallone
Grand Prix of Monza	Formula Two	Fangio
24 Hours of Le Mans	Sports	Chinetti
Grand Prix of Switzerland	Formula One	Ascari
Spa 24 Hours	Sports	Chinetti
Rheims	Formula Two	Ascari
Dolomite Cup	Sports	Vallone
Grand Prix of Holland	Formula One	Villoresi
Grand Prix of Silverstone	Formula One	Ascari
Grand Prix of Europe-Monza	Formula One	Ascari
Grand Prix of Czechoslovakia	Formula One	Whitehead

The Formula One races were won by the 1.5-liter supercharged cars, but in the absence of any competition from Alfa Romeo which had decided to withdraw for one year after three of their drivers, Wimille, Varzi and Count Trossi, were killed. Nevertheless, these were not hollow victories as both Maserati and Talbot gave the Ferraris a good fight. Maserati won the British Grand Prix with de Graffenried at the wheel and Talbot scored victories at the Belgian Grand Prix (Rosier) and the French Grand Prix (Chiron).

It was in Formula Two racing that Ferrari's 1949 record was especially impressive. The cars won every race in which they were represented: Monza, Rome, Bari, Rheims, Brussels, Luxembourg and Naples. By year's end, Ferrari had also won every important sports car race and firmly established his modest Maranello factory as the outstanding producer of sports cars in the world. The most important win was perhaps Luigi Chinetti's heroic performance at the 1949 Le Mans 24 Hour race, proving the true reliability of the seemingly complex Ferrari V-12 engine.

Two Tipo 166 Mille Miglia two-liter barchettas had been brought to Le Mans. The winning car had a rather special engine fitted with needle bearings on the connecting rod journals in place of the normal thin wall insert bearings. Colombo already had experience with this arrangement—which allowed higher revs—on the Tipo 158 Alfa Romeo. In fact, the plain bearing car was actually faster but its driver, Pierre-Louis Dreyfus, crashed it early on in the race. Chinetti's winning car was Chassis Number 0008M, Engine Number 0006M. The numbers have been verified by the records of the Automobile Club de L'Ouest and by the fact that this engine has been examined in recent years and does indeed have needle roller bearings on the connecting rod journals! This historic barchetta has been totally restored and is in the collection of English *Ferrarista* Anthony Bamford.

For himself, Ferrari had only one major challenge left to prove his worth as a serious car constructor: he must beat Alfa Romeo's Tipo 159 Grand Prix car. The fact that he had supervised the creation of this very Alfa was an ironic twist that made the challenge even more enticing. And he met the challenge as he was to meet so many others, with success. ✥

Moresi in a 125 wins the Dutch GP.

CHAPTER 5

Sound & Fury -Ferrari Engines

by Jan Norbye

Jan Norbye has been writing about cars since he was a high school student in his native Norway. Since then, he has specialized in automotive technical writing and automotive history, and his wide-ranging career has taken him to England and HRG, to France with Esso-Standard, and to Sweden where he worked as a Volvo service representative. In 1961, he moved to the United States to become technical editor of Car and Driver *magazine and also became a contributor to other automotive publications including* AUTOMOBILE Quarterly. *He later became automotive editor of* Popular Science Monthly, *a position he recently left to become international editor of* Automotive News *and to do freelance automotive writing. His books include* Sports Car Suspension, The New Fiat Guide, *and* The Wankel Engine. *In this chapter, he describes many of the most important Ferrari engines and explores the engineering concepts which determined their design.*

It was George Mikes (pronounced Mee-kesh), the Hungarian-born humorist, who wrote of his adopted country that: "In England, everything is a compromise. For instance, Yorkshire pudding is a compromise between a pudding and the county of Yorkshire." (From *How to be an Alien,* London, 1947.) What is true of England is just as true of the world at large. We make compromises in our private lives and in our professions. Perhaps no profession is more compromise-ridden than the engineering vocation.

And automotive engineers, be they British, American or Italian, or

whatever, must accommodate opposing factors every step of the way from the initial concept to the final product. Setting the specifications for a new engine, for instance, becomes a matter of trading off displacement against weight, complexity against reliability, stress levels against longevity, better materials against cost. In Detroit, of course, *everything* is compromised by cost.

Other compromises are made in Stuttgart, Coventry, Tokyo, Torino and Billancourt. Compromises are necessary. There is no way to avoid them, faced as we are with the realities of the modern industrial world. The only choice left to the men who have the power of decision-making over the products of an automobile manufacturer is what can be compromised, and how much.

Mass-produced cars resemble each other because the engineers who were responsible for them made the same compromises. At the opposite end of the market, the products have more variety because the designers set different priorities. A Rolls-Royce Corniche and a Ferrari Berlinetta BB may have similar price tags, but the cars could not be more different in appearance, character, purpose and potency. While Rolls-Royce will compromise everything but quality, Ferrari will compromise anything except performance.

Since the philosophy behind Ferrari's cars is based on racing requirements, the first Ferraris were racing cars, and they were built with one purpose: winning. Cost was immaterial, within the scope of Ferrari's activity at the outset. Later, the price of the cars that were sold was simply calculated so as to make sure the racing costs were recovered. This could be done only if Ferrari racing cars did win, and only as long as they continued to win.

The role of Enzo Ferrari himself must not be underestimated when considering the basics of early Ferrari engines. The decision to make a V-12 was his and his alone. Without any formal training as an engineer, Ferrari knew more about race car engineering than many a holder of impressive degrees and diplomas from reputable institutions. He knew the Alfa Romeo engines intimately, and had firm opinions about what it would take to beat them. Alfa's Tipo 158 engine was a long-stroke straight-eight. Ferrari wanted superiority in two areas: piston area and rpm. His reasoning led him to a short-stroke V-12. How could these twelve cylinders give greater piston area and higher rpm? And why would that enable Ferrari to beat Alfa?

The explanation is relatively simple. Formula One engines of the 1945-1951 era were limited to 1500 cc (91 cubic inches) if supercharged and 4500 cc (273 cubic inches) with natural aspiration. Alfa had gone the route of the supercharger—Ferrari did the same.

The Alfetta had eight cylinders of 58 mm bore and 70 mm stroke. That represents a bore/stroke ratio of .828. and a piston area of 211.36 square centimeters (32.76 square inches). Larger piston area can be

obtained in two ways: either by raising the bore/stroke ratio or by increasing the number of cylinders.

Piston area should ideally be as large as possible, since that is the surface the expansion pressure from each individual combustion in each cylinder is active upon. The piston incidentally does not produce positive torque during all 180 degrees of crankshaft rotation that is called the power stroke. The greatest effect is achieved in the first 75 degrees or so, and the expansion forces are spent at the end of 120 degrees. With larger cylinder bore, a larger proportion of the expansion pressure imparts energy to the piston than is the case with small-diameter pistons. Engines with large piston areas tend to have higher absolute torque, and to enjoy superior acceleration in comparison with engines of similar displacement and similar power output but having smaller piston area.

The Alfa engine could have been redesigned with a new 1.00 bore/stroke ratio. That would have given a bore and stroke of 62 mm, and an increase in piston area from 211.36 to 241.5 square centimeters (37.44 square inches) or a piston area increase of 14.3 percent. Alternatively, it could have been redesigned as a V-12 with the same bore/stroke ratio. This would give a bore of 50.7 mm and a stroke of 51.4 mm, with a piston area of 242.25 square centimeters (37.54 square inches) or an increase of 14.6 percent. The two possibilities—new dimensions and more cylinders—could, of course, be combined. A 1500 cc V-12 with a 1.00 bore/stroke ratio would have a 54.2 mm bore and stroke, giving 276.85 square centimeters (42.91 square inches) piston area, or an increase of 31.0 percent over the Alfa engine.

But what Gioacchino Colombo did for Ferrari was even more spectacular. He chose ''over-square'' cylinders with a bore/stroke ratio of 1.047 (55 mm bore and 52.5 mm stroke), giving 285.1 square centimeters (44.2 square inches) piston area—or a 35.0 percent superiority over the Alfetta.

The choice of an ultra-short stroke gave the Ferrari V-12 a number of further advantages, such as higher rpm potential, reduced bearing loads, reduced friction losses, improved crankcase rigidity and a lighter and lower cylinder block. The latter permitted a lower hood line and lowered the engine's center of gravity.

Now for the question of engine rotational speed. It is generally true, though up to a limit, that running an engine faster will produce more power. The limit can be set by the engine's breathing characteristics, or by its mechanical makeup. It is also generally true that higher rpm means higher internal stress levels in the engine. How important is stress in a racing engine? You might think it has little significance, since races are short, and engines can easily be rebuilt between races. But the fact is that a racing engine runs close to maximum stress levels all its active life, yet if it is ever pushed beyond, the result is total failure. In a race, such failure means losing. To Ferrari, high internal stress was a threat to his objective—winning. He also knew that his car could not win unless it had more power than the Alfetta.

The Alfetta produced 275 hp at 8000 rpm in its 1947-1948 stage of development. And it was reliable—extremely reliable—despite running with extreme internal stresses. How do we measure these stresses? Experts have determined that there are two criteria: mean piston speed and piston acceleration.

Mean piston speed is the mean rubbing speed of the piston skirt against the cylinder wall. The piston never travels at a steady speed, of course, it is always speeding up or slowing down. It has to stop and restart at top and bottom dead centers, and reaches peak speed about the halfway point in the stroke. It was generally held for many years that no engine with a mean piston speed in excess of 4000 ft/min (20.32 meters per second) at peak crankshaft speed could be made consistently reliable, because of the high inertia loads on the crankshaft and high stress levels on all moving parts. At 8000 rpm the Alfetta had a mean piston speed of 3675 ft/min (18.667 meters per second). At the end of its racing career it routinely reached speeds of 10,000 rpm—it developed peak power at 9300—which translates into a maximum mean piston speed of 4593 ft/min (23.33 meters per second).

The short-stroke Ferrari was designed for lower piston speed at the same rpm. The 1948 version (with single-stage blower) developed peak power at 7500 rpm which corresponds to 2584 ft/min (13.12 meters per second) piston speed, or nearly 44 percent lower than in the Alfetta.

Piston acceleration is another measure of internal stress levels and inertia loads, and equally valid, since all inertia loads on the piston, wrist pin, connecting rod and crankshaft are dependent upon piston acceleration. The strange fact about piston acceleration is that it has no direct relationship to mean piston speed. What's generally true is that a longer stroke engine reaches higher piston speeds but lower piston acceleration. The short-stroke engine runs with low mean piston speeds, but high piston acceleration.

J. L. Hepworth of the piston-manufacturing firm of Hepworth & Grandage Ltd. has pointed out that experience shows operation at piston acceleration rates in excess of 100,000 ft/sec/sec is known to bring problems. At 8000 rpm the Alfetta was right at this limit, while at 7500 the Ferrari did not exceed 70,000 ft/sec/sec!

By both measures, we can see that the Tipo 158 was running close to or above the theoretical limits, while the Ferrari had wide safety margins, and a vast potential for development. That potential was

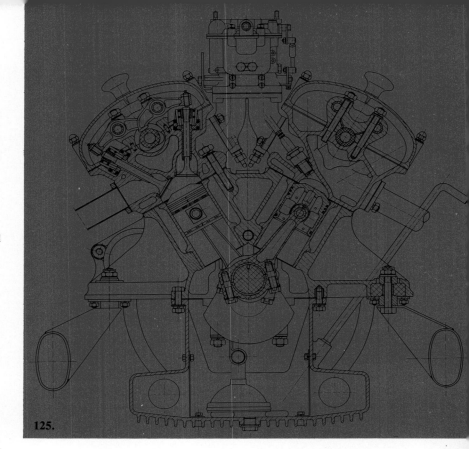

125.

340 America.

fortunate, because the 1948 single-stage version of the Ferrari Tipo 125 Grand Prix engine developed only 230 hp at 7000 rpm, and was clearly no match for the Alfetta.

Had Colombo miscalculated? Had Ferrari given him erroneous instructions? There must have been moments of doubt in Maranello. With more cylinders, there are higher friction losses. The number of valve gear parts increase drastically, and that means added inertia. Compared with Alfa's straight-eight, the V-12 was shorter but certainly had no weight advantage. These factors might have led Ferrari to reconsider his decision to build a V-12.

But no, Ferrari decided the V-12 concept was right. It was only that Colombo had not yet taken full advantage of the opportunities to raise power output that were inherent in the basic design. For instance, he had used only a single camshaft per bank, while the Alfetta was a twin-cam design. Chain camshaft drive was used on the Ferrari V-12, while the Alfetta camshafts were driven by a train of spur gears at the front of the engines. The Ferrari intake ports were siamesed, while the Alfetta had individual ports. The Ferrari's single-stage Roots blower gave about 1.6 Atü pressure while the Alfa—also with a single-stage blower at the time—reached about 1.9 Atü pressure.

The course of development was clear and substantially abetted by the return to Ferrari of the thirty-two year old Aurelio Lampredi. A twin-cam head for the Tipo 125 engine was now designed, eliminating the rocker arms. The cams did not operate the valve lifters directly, reducing valve gear inertia to a bare minimum, but instead used finger followers (which offered greater ease of valve clearance adjustment).

Next, supercharger boost was raised. In theory this was simple—just add a second stage to the Roots blower. In practice, multi-stage supercharging tends to increase the installation space requirement for the engine, it also adds to engine weight, and the blower drive demands more power from the engine.

The four-cam version of the V-12—with two-stage Roots supercharger—retained chain drive for the camshafts, and two valves per cylinder. The two-stage blower built up 2.4 Atü pressure, and the engine delivered 310 hp at 7500 rpm when it made its first race appearance in the Grand Prix of Europe at Monza in September, 1949. But by that time, Alfa Romeo had also gone to two-stage supercharging, and their engine pulled about 335 hp on the dynamometer. In Grand Prix racing Ferrari was still only second best.

In sports car racing during 1949, however, Ferrari exploded on the scene with a victory at Le Mans in June, 1949. The engine—an unsupercharged two-liter—was known as the Tipo 166 (166 cc per cylinder, just as the Grand Prix car had 125 cc per cylinder).

Historically, the 166 was a more significant development than the

125. Ferrari also produced unsupercharged versions of the 125 and used them in sports cars, but the 166 was the first engine that demonstrated the Ferrari wizardry that was going to set the standard for performance in all areas of motor racing for many years to come. In Mille Miglia trim, the 166 V-12 developed 160 hp at 7200 rpm, with a 9.5 compression ratio and three twin-throat Weber carburetors. Bore had been increased to 60 mm and stroke stretched to 58.8 mm, giving 339.3 square centimeter (52.58 square inches) piston area. Stress levels were even lower than in the Grand Prix engine.

In common with its smaller predecessor, the 166 engine had a single-cam cylinder head with hemispherical combustion chambers. Valves were disposed at 45 degrees from the cylinder center line and closed by double hairpin springs. Hairpin springs were chosen because they need little space and have a low rate (but high preload) so as to apply an almost constant closing force on the valve stem.

The camshaft was centrally located in each cylinder head, and opened the valves at each side via rocker arms. Camshafts were driven by triple roller chain. The 125 engine had detachable cylinder heads, with the liners anchored in the block. Because of the high cylinder pressures, the engine suffered from leaks between block and cylinder head. On the 166 engine, the cylinder head was cast integrally with the water jackets.

A typical 166-195-212 series engine.

While Colombo had laid down the overall design of the 166 engine, it was Lampredi who did the detail work. And Lampredi—perhaps the most versatile and creative of Ferrari's engine specialists—followed it up with a series of enlarged versions, such as the 195 Inter, 212 Inter, 340 America and 250 Export, 225 Sport and 250 GT. These engines went into production models, Grand Touring cars and semi-racing sports cars. All of these retained the same single-cam cylinder heads and all these engines came off the drawing board faster than the factory could build them. Lampredi reached a productive peak in this period soon after Colombo ceased his role as Ferrari's engineering consultant in 1949.

He chose the unsupercharged 4.5-liter engine as Ferrari's weapon for a renewed challenge to Alfa Romeo, a decision impossible to defend on the basis of all the evidence from racing experience. The only facts that supported the unblown approach was an occasional victory by Talbot, easily attributed to its stamina and low fuel consumption as opposed to its power and agility.

But Lampredi decided that the supercharged engine would be defeated by its drawbacks—complexity, bulk, weight, high fuel consumption, and loss of power to the blower drive itself, (all of which also lessened its appeal for daily street use) and importantly, an unjust displacement limitation. The three-to-one displacement ratio was

not fixed by laws of nature. It had been fixed by the FIA committee that had proposed the 1938 Grand Prix formula, the formula still in effect after the war.

In 1950 the best 1.5-liter supercharged Grand Prix engines developed about 220-235 hp per liter. To equal that in terms of absolute power, Lampredi only needed to extract 70-80 hp per liter with an unblown 4.5-liter. But Lampredi was already thinking in terms of 100 hp per liter with unsupercharged engines of simpler design and lower stress levels than their blown counterparts. It took remarkably little time to prove how right he was.

The first unsupercharged V-12 Ferrari racing cars were the Formula Two Tipo 166 models built in 1949 and raced with great success. These cars were the starting point for the next generation of Formula One racers. Lampredi did not suddenly drop a 4.5-liter engine into a chassis designed for a two-liter, however. It was a gradual process, and the chassis was modified to accept successively larger engines and higher torque, step by step.

First the 212 engine—displacing 2560 cc (156.16 cubic inches) was used, followed by the 275 of 3322 cc (202.65 cubic inches). With the 340 Mexico came a 4104 cc (250.22 cubic inches) and finally—in the fall of 1950—the 375 Grand Prix powerplant. This 4494 cc (274.4 cubic inches) engine was finished on time to contest the Italian Grand

Partially stripped 166-195-212.

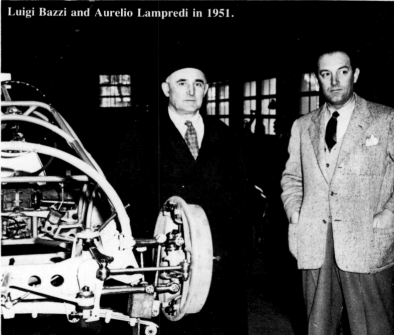

Luigi Bazzi and Aurelio Lampredi in 1951.

A GALLERY OF FERRARI
FOURS:
the enormously successful 500
F2 (above), exhaust side of a
500 Modial (above right), and
two views of the famous 500
Testa Rossa engine.

Prix at Monza and though its 330 hp was some 35 hp down on the Alfetta, Ascari managed to lead the race for a time before his crankshaft broke—because of a faulty billet—and he took over a teammate's car to finish in second place. Alfa Romeo, having now won the World Championship, retired on their laurels leaving Ferraris to finish 1-2-3 at the Spanish Grand Prix. Incidentally, this engine firmly established a Ferrari tradition of using Vandervell thin shell bearings—which had earlier replaced thicker babbitted-style bearings in 125 and 166 engines—to support the crankshaft. About these bearings, it might be said that Ferrari was benefitting from U.S. bearing technology since Vandervell was a licensee of the Cleveland firm Clevite, the bearings' developer.

The advent of the 1951 season saw Alfa Romeo with engines delivering between 405 and 425 hp and improvements in the Ferrari engine were obviously called for. Still Lampredi preferred not to design a twin cam head or use multiple valves, and preferred to keep the engine as simple as possible. Thus the main change he effected for 1951 was a new twin plug head which he and Luigi Bazzi anxiously hoped would be enough to insure a substantial horsepower gain. To the great joy of the Ferrari engineers, dynamometer tests of the engine showed a full 10 hp gain over its predecessor, 370 hp at 7500 rpm.

If the engine remained less powerful than the Alfa's, it also ran at lower stress levels than the Alfa Romeo Tipo 159—Alfa's designation for its de Dion rear suspension 1951 GP car—consumed less fuel and proved astonishingly reliable. Engine failures accounted for only two DNF's out of some thirty-eight starts for the 375 and it finally powered Froilan Gonzalez to victory over Alfa at Silverstone. Alfa retired from Grand Prix racing after the 1951 season but Ferrari, of course, moved ahead with unflagging ardor and looked to the interim two-liter displacement class that would comprise Formula One until the new 2.5-liter formula became official in 1954.

The Ferrari 166 engine might have been perfect for the interim formula but Lampredi was anxious to try a new powerplant, the drawings for which he had completed during the winter of 1950-51. His successor to the lovely little V-12 was, of all things, an in-line four. Yes, the same basic layout that graced agricultural tractors, Fiat Balillas and any number of other sedans and sports cars. But there were important differences between these everyday motive sources and the new Ferrari engine.

The five-bearing Ferrari four introduced a new type of valve gear to the marque—valves with dual coil springs mounted concentrically with the valve stem in addition to two hairpin springs on the side. The coils, of very thin wire, were used to load the lifter assembly against the cam—there were twin overhead camshafts, gear driven and acting upon inverted-cup type lifters—while the hairpin springs took care of valve closing. The lifters had rollers which contacted the cam lobes. The valves were disposed at a narrow 58-degree included angle in the twin plug head, a light alloy casting with screwed-in liners like all of Lampredi's V-12's.

This engine was uniquely Aurelio Lampredi's own. Although the engineer knew that a four-cylinder two-liter engine's piston area would be far smaller than a similar displacement V-12's, and that the stroke would be much longer, he did not give the same importance to piston area and piston speed/acceleration as had both Colombo and Ferrari. Lampredi had his own priorities. He knew—for one thing—that he could save weight by using a simple four-cylinder block, and the 500 F2 weighed 348 pounds compared to 440 pounds for the 166 V-12. He estimated he could raise the power/weight ratio by 12-15 percent compared with the earlier engine. Further, the number of moving parts was reduced by some 65 percent.

The first four-cylinder—built in the spring of 1951—developed 170 hp against the V-12's 155 hp. While the 166 boasted a power/weight ratio of .352 hp per pound and 77.8 hp per liter, the new four gave .488 hp per pound and 85.86 hp per liter. Lampredi had outdone his own estimations for the power/weight ratio: the new engine's was 27.9 percent improved over the V-12. There was a 9.4 percent increase in specific power output.

The Ferrari 500 F2 proved an unbeatable car and powered Alberto Ascari to the World Championship in both 1952 and 1953. By then the four—now with the angle between valves widened to 100 degrees to improve breathing and give a lower engine—had spawned a new generation of Ferrari sports cars beginning with the two-liter 170 hp 500 Mondial. A later version of the four—tuned to 180 hp at 7000 rpm—was to become the first Testa Rossa. In this powerplant, the dual coil lifter springs were replaced by a single coil and the engine's success led Lampredi to eventually increase the displacement—in eleven increments—from 1985 cc (121.08 cubic inches) to a full three liters (183 cubic inches) in the 250 hp 750 Monza. Still, most of these engines displayed great reliability.

Lampredi's continued success with the four-cylinder layout convinced him to retain it for the 1954 Grand Prix season. He designed the Tipo 625 for the new 2.5-liter formula, with 94 mm bore and 90 mm stroke. The engine now delivered 230 hp at 7000 rpm, and by season's end—reached 245 hp. Even the higher figure was not sufficient to seriously challenge the new Maserati 250F—designed by Colombo—or the magnificent Mercedes Benz W196 with its 256 hp. Neither was the subsequent introduction—in 1955—of the Ferrari 555 Super Squalo with a 270 hp four enough to upset the balance. By then, the Mercedes-Benz developed 280 hp. Both the 1954 and 1955 seasons were unsuccessful for Ferrari Formula One cars.

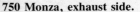

750 Monza, exhaust side.

750 Monza showing Weber 58 DCOE carburetors.

1957 Formula One Ferrari/Lancia V-8.

It was not until 1956 that Ferrari's Formula One fortunes improved. At the end of the 1955 season, Lancia withdrew from its financially ruinous and tragic racing program and donated all its cars, spare parts, tools and drawings to Ferrari and even offered Ferrari the services of the brilliant engineer who had been responsible for the GP Lancia, Vittorio Jano.

Thus Ferrari suddenly had a V-8 Grand Prix car for the 1956 season, and a number of hybrid Lancia-Ferrari cars were tried out with various cylinder dimensions until Jano and Bazzi settled on a 76 mm bore and 68.5 mm stroke yielding 2487 cc (151 cubic inches). This V-8 delivered 275 hp at 8000 rpm and powered Juan Fangio to another World Championship in 1956.

This engine—called the D50 by Lancia—was derived from both the Aurelia family of V-6 sports car engines and a subsequent series of Lancia racing cars, the D20-D24. It had four chain-driven overhead camshafts and detachable cylinder heads. But once it became part of the Ferrari organization, the engine underwent several changes eventually leading to the Dino.

In 1957, Ferrari Formula One cars continued to use the ex-Lancia V-8 engine, now revised with 80 mm bore and 62 mm stroke and tuned to deliver 285 hp at 8800 rpm. The engine had other changes as well, engineered by Carlo Chiti, a thirty-three year old who joined Ferrari from Alfa Romeo where he had worked in the racing department as well as on experimental and production engine designs. A Tuscan with a diploma from the University of Pisa where he studied aeronautical engineering, Chiti—who became Ferrari's chief engineer—would introduce new ideas both in Ferrari engines and chassis. During 1957, he tried both three and four valve cylinder heads and began experiments with Lucas fuel injection.

Meanwhile development of the 375 V-12 had proceeded and the engine was enlarged in 1954 to become the 375 Plus with the 375's bore but a stroke lengthened from 68 mm to 74.5 mm, giving 4954 cc (302.2 cubic inches). Both the 375 and the 375 Plus were rated at the same 340 hp but the 375 Plus developed its peak power at 6500 rpm while the 375 reached peak power at 6800 rpm. The 375 Plus powered a Ferrari to victory at Le Mans in 1954 and the engine was used in a number of different 375 and 410 SA production models, mostly heavy coupes, and enjoyed a long production life by Ferrari standards although the volume was small.

With Lampredi's departure at the end of the 1955 season, Ferrari began bringing in more outside consultants to lead his engineering staff and hired two of the men who had worked on the 815 project, Vittorio Bellentani and Alberto Massimino. As it turned out, Massimino was mostly assigned to chassis development and

130 S, 24 plugs, four distributors.

340 or 375 competition engine with dual magnetos.

modifications, and as a result plays no direct part in this story. But Bellentani was assigned to engine development, where he worked closely with Andrea Fraschetti. Out of that collaboration came not only the largest four-cylinder Ferrari ever built but also the first four-cam V-12 since the abortive 125 GP of 1949.

Vittorio Bellentani was born in 1906, studied engineering in Switzerland, and first gained attention when he designed the Mignon motorcycle for the Gueroni works in the early Thirties. He spent the war years with Ferrari in Modena, and served as technical director of Maserati from 1946 to 1952. His approach to engine design at Ferrari was to evaluate what was at hand, discard what was poor or indifferent, use what was good, and try to improve it if possible.

The four-cylinder Tipo 860 appeared in 1956. The cylinder dimensions were ''undersquare'' with 102 mm bore and 105 mm stroke, giving 3438 cc (209.1 cubic inches). It delivered 310 hp at 6200 rpm, and powered Ferraris to 1st and 2nd at Sebring in 1956.

The 3.5-liter V-12 delivered twenty more horsepower with lower internal stress. This engine was known as the 290 Mille Miglia. Its four overhead camshafts were chain-driven from the front of the crankshaft. The valves were disposed at an included angle of 60 degrees. With a bore of 73 mm, Bellentani and Fraschetti were able to go as high as 35 mm on inlet valve head diameter and 29 mm for the exhaust valve head. With a stroke of 69.5 mm, mean piston speeds and piston acceleration rates were moderate, which meant that the engine had a great reliability/durability potential. As a result it was

Four cam V-12 racing engi

chosen for the long-distance endurance events, making its debut in the 1956 Tour of Sicily.

During an intensive tuning and development process Bazzi coaxed ever more power out of it. Soon it was giving 320 hp at 6800 rpm, and eventually it delivered 350 hp at 7200 rpm with a peak torque of 32 kgm (231.4 lb/ft) at 5200 rpm. The engine was enlarged to 3.8-liters (232 cubic inches) in time for the Sebring twelve-hour race in 1957, and its displacement was increased further to 4023 cc (254.4 cubic inches)—the engine was now designated 312 LM—for Le Mans.

The Le Mans effort that year began on a low note. Engineer Fraschetti, who had contributed much to Ferrari including the six-cylinder engines recounted elsewhere in this book, was killed in a road accident in Italy while testing a new prototype. The race itself was unsuccessful for Ferrari despite Mike Hawthorn's lap record which was to stand until 1962. Hawthorn's 4.1-liter car blew up; Jaguars finished in the top four places, and the 3.8-liter Ferrari of Stuart Lewis-Evans and co-driver Severi finished fifth.

By 1957, Vittorio Jano was well along with the development of his V-6 Ferrari—the Dino—a configuration which had originally been suggested by Ferrari's son. Although the V-6 would eventually find its way into Ferrari road cars, it began life as a Formula Two engine—the 156 F2—for the 1957 season. Although Jano had used a conventional 60-degree layout for his Lancias, he now chose a 65-degree layout for what was to become the Formula One Ferrari Tipo 156. Why 65

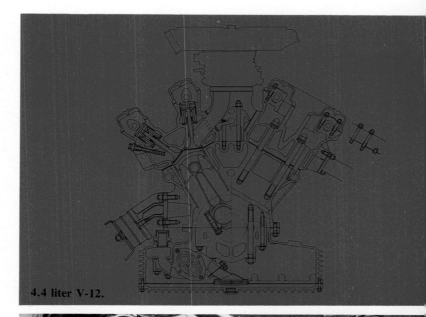

4.4 liter V-12.

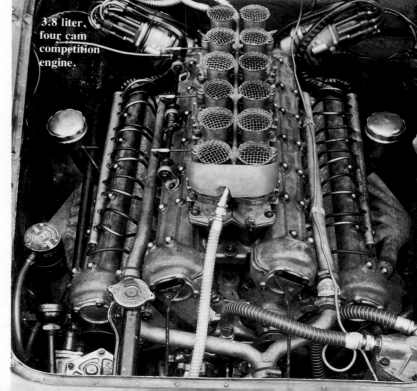

3.8 liter, four cam competition engine.

n the 290, 315 S and 335 S.

Early 250 GT with front-mounted distributors.

A dry sump 250 TR with outside plug location.

degrees? Simply because he needed more space between the banks to design intake ducts that would assure proper breathing. But what about the crankshaft—and engine balance? No problems for an old technical twister like Jano.

To understand the crank arrangement in the Dino V-6, you should stop thinking about it as two banks of three cylinders and instead approach it as three V-twins in line. How did Jano obtain normal 120-degree phasing between firing impulses? By spacing some crankpins at 55 degrees and others at 185 (55 + 65 = 120; 185 − 65 = 120) instead of the usual 60-180 spacing. Look at the front V-twin as an example of how this works. These two cylinders should fire 120 degrees apart, then the middle V-twin, and finally the rear V-twin. Looking at the engine head-on, the right side fires first. That means that when cylinder No. 1 is fired, cylinder No. 2 should be 120 degrees of crankshaft rotation from firing.

As the right bank has a 65-degree advance built into it because of the V-angle, only 55 degrees remains for actual crankpin spacing. Next we must consider the proper phasing between each of our V-twins. The right cylinder in the middle V-twin should fire 120 degrees after the left cylinder in the foremost pair of cylinders, thus at 185 degrees. The engine fires six times at equal intervals in two crankshaft revolutions.

Initially the 156 engines developed 170 hp at 8500 rpm. By 1958 the engine had been enlarged to 2.4-liters and developed 280 hp at 8500 rpm. Now called the Dino 246, it powered Mike Hawthorn to

From a solid billet of steel— a 250 GT crank.

250 GT engine for a Boano coupe, note off-set gearshift.

500 TRC.

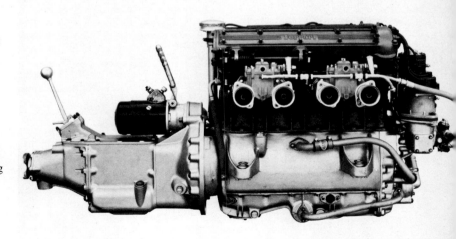

the World Drivers Championship that year and—as the 290 hp Dino 256—proved a competitive and durable engine throughout the remaining two seasons of the 2.5-liter Formula.

Incidentally, with the arrival of the Dino engines, Ferrari introduced a new system of engine designations. Instead of simply identifying the engine according to the displacement of a single cylinder and leaving the number of cylinders out of the designation, the new system used a different three-digit combination. In it, the first two numbers indicated total displacement in deciliters and the third numeral indicated the number of cylinders. Thus the Formula Two Dino engine—the 156—derived from its 1.5-liter displacement and 6 cylinders. (A third numbering system introduced years earlier by Lampredi and used only within the factory was simply chronological.)

While the V-6 and V-8 engine programs matured, Chiti continued development of the V-12 with the assistance of other Ferrari engineers including Franco Rocchi and Giorgio Salvarani. Much of their thinking about V-12 development was governed by the 1958 sports car prototype regulations limiting such cars to three-liters displacement. Since Chiti did not believe that twin-cam heads would be practical in this displacement because of their added weight and complication, he reverted to the single-cam cylinder head with rocker arms and hairpin-type valve springs. Many of the Colombo V-12's had used siamesed intake ports, but Chiti now redesigned the cylinder head with larger individual intake ports, adding considerably to the engine's volumetric efficiency.

For the cylinder block, Chiti revived the 250 GT engine of 1953 with its 73 x 58.8 mm dimensions, a powerplant based directly on the 166 V-12. Like that engine, however, Chiti used detachable heads with liners shrunk into the block. The 250 TR was the first Ferrari to use conventional double-coil valve springs, these parts imported from the U.S. With its new cylinder head—colored red—the 250 engine was dubbed Testa Rossa and it provided Ferrari with several seasons of victories in sports car races during a period when their Grand Prix cars were failing.

The 250 Testa Rossa of 1958, as raced at Le Mans, delivered at least 300 hp at 7200 rpm. Yet it was one of the lowest-stressed engines in the race, which accounts for its amazing staying power. It became the backbone of the production car series as well as the sports car prototypes until 1964. In fact, its successors in the GT cars are its direct derivatives. After more than ten years of experiments and developments, Ferrari had returned to almost the same concept with which he had begun: the road and racing engines were basically the same.

Meanwhile, Grand Prix racing cars had—by 1960—undergone a basic change of architecture, since the engine had been moved from the front of the chassis to a position behind the driver. This position offered a wider berth for the engine and an opportunity to lower the engine's center of gravity. Accordingly, Chiti responded by folding down the 65-degree Dino 156 to a 120-degree V-angle. This idea was not new, and to give credit where credit is due, it is part of the record that Lampredi had designed a 120-degree V-6 back in 1950-1951. Chiti's revised Dino 156 was very successful, substantially more powerful (190 hp at 9400 rpm) than its competition, and Phil Hill won the 1961 World Championship in a series of races that in retrospect seem more like demonstration runs.

30 GT.

Racing Dino.

Conscious of new V-8 Grand Prix engines being prepared in Great Britain by BRM and Coventry Climax, Chiti was never satisfied with the Ferrari V-6's power superiority, and he tried a number of different cylinder heads on the same basic engine. In chronological order, they were

1. Two valves per cylinder with two spark plugs.
2. Two valves per cylinder with three spark plugs.
3. Three valves per cylinder with two spark plugs.
4. Four valves per cylinder with one spark plug.

Progress was slow with the new heads and the first version was retained for the 1962 season, after Chiti left for ATS. It had the same bore and stroke as the 250 Testa Rossa.

Though much of his staff had departed with Chiti, Ferrari was not tempted to call in outside consultants. He had two young turks in the design office, and he told each of them to design a Formula One engine. Angelo Bellei designed a four-cam V-8, and Mauro Forghieri came up with a horizontally opposed four-cam twelve-cylinder unit.

Both designs being new and untried, Ferrari decided to rely on the 120-degree Dino 156 for the 1963 season. But the V-8 and flat-12 proposals were built, tested and evaluated in preparation for 1964.

The V-8 reached maturity first, and powered Surtees to the World Championship for Ferrari in 1964. The flat-12 was thrown into the fray in 1965, but could not repeat the V-8's accomplishments. Both these engines were out-and-out racing designs with gear-driven camshafts and both were built with two-, three-, and four-valve heads and a multiplicity of spark plug combinations. The flat-12 may have had a lasting influence on Ferrari engine design, but the growing use of that configuration by Ferrari in later years may also stem from the fact that Forghieri was made chief engineer of the racing section, while Bellei was divorced from the racing program and appointed chief production car engineer.

Production Dino.

 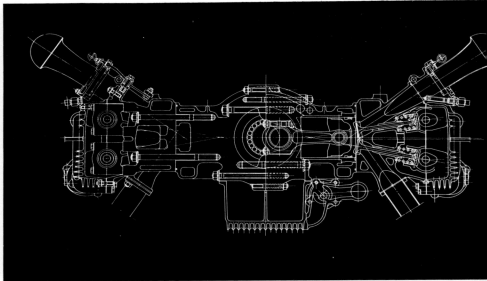

1970 512 S.   312 B.

When Enzo Ferrari learned of the three-liter (183 cubic inches) GP formula adopted for 1966, he was understandably pleased. Naturally, the 250 GT could not be turned into a Formula One engine, but the factory's experience with three-liter V-12's was bound to be helpful.

Forghieri designed a four-cam V-12 known as the 312. It was a 60-degree engine with 77 x 53.5 mm bore and stroke, delivering 360 hp at 10,000 rpm. This engine was fed with Lucas fuel injection, had two plugs per cylinder and, initially, a two-valve head. In 1967 a three-valve head was tried and then, for Monza, a four-valve head was introduced. None of these developments were enough to make the 312 a consistent force in GP racing during 1966 and 1967 and Ferrari had little success with its Formula One program during the late Sixties.

In an effort to improve things, the thirty-three-year-old Forghieri, assisted by a younger engineer named Caliri and Giancarlo Bussi, flattened the 312 to a horizontally opposed unit for the 1969 season, naming it the 312B. This engine actually grew out of a twelve-cylinder two-liter boxermotor prototype engine designed by a young engineer named Jacaponi that was given the designation 212E. It gained the Mountain Championship for Ferrari in 1969; however, at the end of that season, Jacaponi left Ferrari to join Abarth.

The fuel-injected 312B developed 436 hp at 11,000 rpm. Its cylinder dimensions were highly oversquare with 78.5 mm bore and

51.5 mm stroke, and the four valves per cylinder were canted at the smallest angle of any Formula One engine—25 degrees—after much experimentation. The final angle was chosen on the basis of mechanical convenience and volumetric efficiency.

The 312B's crankshaft—machined from a $375 solid billet of U.S.-made steel—runs in four main bearings while the production model V-12 crankshafts rest on seven mains. Forghieri tried every possible configuration from three to seven bearings in the GP powerplant but decided against seven, because it makes the engine (a) long, (b) heavy, and (c) adds to friction losses. With three mains, the crankshaft becomes too whippy, revealing alarming torsional deflection angles and thus inducing metal fatigue and threatening the engine's life. The 312 was destined to move Ferrari back into really competitive status in Formula One.

During the 1960's, the Ferrari engine specialists had also worked together to produce a number of striking sports car prototype racing engines. They were 60-degree V-12's in the traditional Ferrari pattern and appear to be little more than enlarged GP units. Rocchi was the principal design engineer on the 330 P4, for instance, a four-cam four-liter (242 cubic inches) for long-distance racing. It appeared in 1967—developing 450 hp at 8000 rpm—and overwhelmed all challengers to win Daytona.

GTB/4.

Year	Model	No. Cylinders & Arrangement	Bore & Stroke	Displacement		Max. RPM	Compression Ratio	hp	Camshaft (per cylinder head)	Number and type Weber Carburetors or fuel injection	Fuel G—Gas M—Mixture
				CC per cylinder	total						
1946	125/GT	12 V 60°	55 x 52.5	124.73	1496.77	5600	8:1	72	One	1-30 DCF	G
1947	125/S	12 V 60°	55 x 52.5	124.73	1496.77	7000	8.5:1	100	One	1-30 DCF	G
1947	159/S	12 V 60°	59 x 58	158.57	1902.84	7000	8.5:1	125	One	1-30 DCF	G
1948	166/S	12 V 60°	60 x 58.8	166.25	1995.02	7000	8.5:1	150	One	1-30 DCF	G
1948	125/F1	12 V 60°	55 x 52.5	124.73	1496.77	7000	6.5:1	230	One	1-50 WCF	M
1948	166/F2	12 V 60°	60 x 58.8	166.25	1995.02	7000	10:1	160	One	3-32 DCF	M
1948	166/Inter	12 V 60°	60 x 58.8	166.25	1995.02	6000	8:1	115	One	1-32 DCF	G
1948	166/MM	12 V 60°	60 x 58.8	166.25	1995.02	6600	8.5:1	140	One	3-32 DCF	G
1949	166/FL	12 V 60°	60 x 58.8	166.25	1995.02	7000	6.5:1	310	One	1-40 DO3C	M
1949	125/F1	12 V 60°	55 x 52.5	124.73	1496.77	8000	6.5:1	280	Two	1-40 DO3C	M
1950	195/S	12 V 60°	65 x 58.8	195.08	2341.02	7000	8.5:1	170	One	1-36 DCF	G
1950	275/S	12 V 60°	72 x 68	276.86	3322.34	7200	8:1	270	One	3-40 DCF	G
1950	275/F1	12 V 60°	72 x 68	276.86	3322.34	7300	10:1	300	One	3-42 DCF	M
1950	340/F1	12 V 60°	80 x 68	341.80	4101.66	7000	12:1	335	One	3-42 DCF	M
1950	375/F1	12 V 60°	80 x 74.5	374.47	4493.73	7000	11:1	350	One	3-42 DCF	M
1951	212/F1	12 V 60°	68 x 58.8	213.54	2562.51	7500	12:1	200	One	3-32 DCF	M
1951	166/F2	12 V 60°	63.5 x 52.5	166.26	1995.16	7200	11:1	160	Two	3-32 DCF	M
1951	212/Export	12 V 60°	68 x 58.8	213.54	2562.51	6500	8:1	150	One	3-32 DCF	M
1951	195/Inter	12 V 60°	65 x 58.8	195.08	2341.02	6000	7.5:1	135	One	1-36 DCF	G
1951	340/America	12 V 60°	80 x 68	341.80	4101.66	6000	8:1	230	One	3-40 DCF	G
1951	375/F1	12 V 60°	80 x 74.5	374.47	4493.73	7300	11:1	360	One	3-46 DCF/3	M
1951	375/F1	12 V 60°	80 x 74.5	374.47	4493.73	7500	12:1	380	One	3-46 DCF/3	M
1951	500/F2	4 in-line	90 x 78	496.21	1984.85	7500	13:1	185	Two	2-45 DOE	M
1951	212/Inter	12 V 60°	68 x 58.8	213.54	2562.51	7000	8:1	160	One	3-32 DCF	G
1952	225/S	12 V 60°	70 x 58.8	226.28	2715.46	7200	8.5:1	210	One	3-36 DCF	

Year	Model	No. Cylinders & Arrangement	Bore & Stroke	Displacement CC per cylinder	Displacement total	Max. RPM	Compression Ratio	hp	Camshaft (per cylinder head)	Number and type Weber Carburetors or fuel injection	Fuel G—Gas M—Mixture
1952	250/S	12 V 60°	73 x 58.8	246.10	2953.21	7500	9:1	230	One	3-36 DCF	G
1952	735/S	4 in-line	102 x 90	735.41	2941.66	6800	9:1	225	Two	2-50 DCOA	G
1952	375/Ind	12 V 60°	80 x 74.5	374.47	4493.73	7500	13:1	400	One	3-40 IF4C	M
1952	340/Mexico	12 V 60°	80 x 68	341.80	4101.66	6600	8:1	280	One	3-40 DCF/3	M
1953	250/EU	12 V 60°	68 x 68	246.95	2963.45	7000	8.5:1	220	One	3-36 DCL/3	G
1953	250/MM	12 V 60°	68 x 68	246.95	2963.45	7200	9:1	240	One	3-36 DCF/3	G
1953	625/F1	4 in-line	94 x 90	624.58	2498.32	7000	13:1	240	Two	2-50 DCOA	M
1953	553/F2	4 in-line	93 x 73.5	499.27	1997.11	7200	13:1	180	Two	2-52 DCOA/3	M
1953	700	4 in-line	99 x 90	692.79	2771.16	6800	12:1	250	Two	2-45 DCOA	M
1953	375/America	12 V 60°	84 x 64	376.91	4522.94	6500	8.4:1	300	One	3-40 DCZ3	G
1953	500/Mondial	4 in-line	90 x 78	496.21	1984.85	7000	8:1	155	Two	2-45 DOE	G
1953	250/I	12 V 60°	68 x 68	246.95	2963.45	6300	7.5:1	505	One	2-40 DCM	M
1954	375/Plus	12 V 60°	84 x 74.5	412.86	4954.34	6000	9.2:1	330	One	3-46 DCF3	G
1954	750/Monza	4 in-line	103 x 90	749.90	2999.62	6400	9:1	260	Two	2-58 DCOA/3	G
1954	625/F1	4 in-line	100 x 79.5	624.39	2497.56	7200	13:1	260	Two	2-50 DCOA/3	M
1954	625/F1	4 in-line	94 x 90	624.58	2498.32	7200	13:1	250	Two	2-50 DCOA/3	M
1954	625/S	4 in-line	94 x 90	624.58	2498.32	7000	9:1	220	Two	2-50 DCOA/3	G
1954	500/Mondial	4 in-line	90 x 78	496.21	1984.85	7200	9:1	165	Two	2-40 DCOA/3	G
1954	250/GT	12 V 60°	73 x 58.8	246.10	2953.21	6600	9:1	200	One	3-36 DCZ/3	G
1954	306/S	6 in-line	90 x 78	496.21	2977.29	6600	8:1	230	Two	3-50 DCOA/3	
1955	256/F1	6 in-line	82.4 x 78	415.94	2495.68	7000	13:1	215	Two	3-50 DCOA/3	M
1955	252/F1	2 in-line	118 x 114	1246.69	2493.38	5200	13:1	160	Two	2-42 DCOD	M
1955	376/S	6 in-line	94 x 90	624.58	3747.48	6400	8:1	280	Two	3-58 DCOA/3	G
1955	446/S	6 in-line	102 x 90	735.41	4412.49	6000	8.5:1	330	Two	3-50 DCOA/3	G
1955	446/Ind	6 in-line	102 x 90	735.41	4412.49	6300	9:1	360	Two	3-50 DCOA/3	G
1955	250/GT	12 V 60°	73 x 58.8	246.10	2953.21	6800	8.5:1	210	One	3-36 DCZ/3	G
1956	354/S	4 in-line	102 x 105	857.98	3431.93	6000	8.5:1	280	Two	2-58 DCOA/3	G

Year	Model	No. Cylinders & Arrangement	Bore & Stroke	Displacement CC per cylinder	Displacement total	Max. RPM	Compression Ratio	hp	Camshaft (per cylinder head)	Number and type Weber Carburetors or fuel injection	Fuel G—Gas M—Mixture
1956	290/MM	12 V 60°	73 x 69.5	290.88	3490.61	7300	9:1	320	One	3-36 IR4/C1	G
1956	500/TR	4 in-line	90 x 78	496.21	1984.85	7400	9.1:1	190	Two	2-40 DCO/3	G
1956	250/S	12 V 60°	73 x 58.8	246.10	2953.21	7500	9:1	240	One	3-40 DCZ	G
1956	410/SA	12 V 60°	88 x 68	413.46	4961.57	6500	8.5:1	340	One	3-46 DCF/3	G
1956	156/F2	6 V 65°	70 x 64.5	248.22	1489.35	9000	10:1	180	Two	3-38 DCN	G
1957	290/S	12 V 60°	73 x 69.5	290.88	3490.61	8000	9:1	330	Two	6-42 DCN	G
1957	196/F1	6 V 65°	77 x 71	330.62	1983.72	8500	9:1	220	Two	3-42 DCN	G
1957	298/S	8 V 90°	81 x 72.2	372.04	2976.37	8000	11.5:1	280	Two	4-40 IDF	G
1957	156/S	6 V 65°	70 x 64.5	248.22	1489.35	8500	9.2:1	170	Two	3-36 DCN	G
1957	315/S	12 V 60°	76 x 69.5	315.28	3783.40	7800	9:1	360	Two	6-42 DCN	G
1957	335/S	12 V 60°	77 x 72	335.27	4023.32	7800	9.2:1	390	Two	6-42 DCN	G
1957	312/S	12 V 60°	73 x 58.8	246.10	2953.21	8200	9.5:1	320	Two	6-38 DCN	G
1957	226/F1	6 V 65°	81 x 71	365.86	2195.17	8500	9.8:1	240	Two	3-42 DCN	G
1958	246/F1	6 V 65°	85 x 71	402.88	2417.33	8500	9.8:1	280	Two	3-42 DCN	G
1958	412/MI	12 V 60°	77 x 72	335.27	4023.32	8500	9.4:1	415	Two	6-42 DCN	G
1958	296/S	6 V 65°	85 x 87	493.68	2962.08	8000	9:1	300	Two	3-46 DCN	G
1958	326/MI	6 V 65°	87 x 90	535.02	3210.12	7500	9:1	330	Two	3-54 DCN	G
1958	156/S	6 V 60°	72 x 64.5	262.61	1575.67	8000	9:1	165	One	3-38 DCN	G
1958	196/S	6 V 60°	77 x 71	330.67	1983.72	7800	9.8:1	195	One	3-42 DCN	G
1958	196/GT	6 V 60°	77 x 71	330.67	1983.72	7500	9:1	175	One	3-38 DCN	G
1958	256/F1	6 V 65°	86 x 71	412.42	2474.54	8800	9.8:1	290	Two	3-42 DCN	G
1958	250/GT	12 V 60°	73 x 58.8	246.10	2953.21	7200	9:1	235	One	3-38 DCN	G
1958	250/TR	12 V 60°	73 x 58.8	246.10	2953.21	7500	9.8:1	290	One	6-38 DCN	G
1959	246/F1	6 V 60°	85 x 71	402.88	2417.33	8000	9.8:1	250	One	3-42 DCN	G
1959	156/F2	6 V 60°	73 x 58.8	246.10	1476.60	8000	9.2:1	150	One	3-38 DCN	G
1959	854	4 in-line	64 x 64	212.37	849.48	7000	7.5:1	68	One	2-38 DCO	G
1959	950	4 in-line	67 x 69	243.26	973.07	7000	9.1:1	80	One	2-32 WX46	G
1960	1000	4 in-line	69 x 69	258.01	1032.04	7200	9.2:1	98	One	2-38 DCO/A3	G
1960	256/F1	6 V 60°	86.4 x 71	416.27	2497.62	7600	11:1	245	One	3-42 DCN	G
1960	276/S	6 V 60°	90 x 71	451.68	2710.09	7500	9.8:1	255	One	3-42 DCN	G
1960	156/F2	6 V 65°	73 x 58.8	246.10	1476.60	9200	9.8:1	185	Two	3-42 DCN	G
1960	250/I	12 V 60°	73 x 58.8	246.10	2953.21	7500	9.2:1	300	One	Inject. Bosch	G
1960	246/I	6 V 60°	85 x 71	402.88	2417.33	8500	9.8:1	280	Two	Inject. Bosch	G
1960	400/SA	12 V 60°	77 x 71	330.62	3967.44	7000	8.8:1	340	One	3-42 DCN	G
1960	250/GT	12 V 60°	73 x 58.8	246.10	2953.21	7400	9.2:1	250	One	3-38 DCN	G
1961	156/F1	6 V 120°	73 x 58.8	246.10	1476.60	9500	9.8:1	190	Two	2-4 IF3C	G
1961	296/S	6 V 65°	87 x 82	487.46	2924.78	7500	9:1	295	Two	3-45 DCN	G
1961	248/GT	8 V 90°	77 x 66	307.33	2458.70	7500	9.2:1	190	One	4-40 IF2C	G
1961	156/F1	6 V 65°	81 x 48.2	249.40	1496.43	10500	9.8:1	200	Two	3-42 DCN	G
1961	156/F1	6 V 65°	67 x 70	246.79	1480.73	9500	9.8:1	185	Two	3-42 DCN	G
1962	196/S	6 V 60°	77 x 71	330.62	1983.72	7800	9.8:1	200	One	3-42 DCN	G
1962	268/S	8 V 90°	77 x 71	330.62	2644.96	7500	9.8:1	260	One	4-40 IF2C	Gas
1962	250/GTO	12 V 60°	73 x 58.8	246.10	2953.21	7400	9.8:1	290	One	6-38 DCN	G
1963	250/P	12 V 60°	73 x 58.8	246.10	2953.21	7800	9.8:1	300	One	6-38 DCN	G
1963	330/GT	12 V 60°	77 x 71	330.62	3967.44	7000	8.8:1	290	One	3-40 DCZ6	G

Year	Model	No. Cylinders & Arrangement	Bore & Stroke	Displacement CC per cylinder	Displacement total	Max. RPM	Compression Ratio	hp	Camshaft (per cylinder head)	Number and type Weber Carburetors or fuel injection	Fuel G—Gas M—Mixture
1963	158/F1	8 V 90°	64 x 57.8	185.94	1487.54	10700	9.8:1	190	Two	4-38 IDM	G
1963	156/F1	6 V 120°	73 x 58.8	246.10	1476.60	10500	9.8:1	205	Two	Inject. Bosch	G
1963	186/GT	6 V 60°	77 x 64	298.02	1788.14	7000	9.2:1	156	One	3-38 DCN	G
1964	500/SA	12 V 60°	88 x 68	413.46	4961.57	6500	9:1	360	One	3-40 DCZ6	G
1964	330/P	12 V 60°	77 x 71	330.62	3967.44	7300	9.8:1	370	One	6-38 DCN	G
1964	158/I	8 V 90°	67 x 52.8	186.15	1489.23	11000	9.8:1	210	Two	Inject. Bosch	G
1964	512/F1	12 V 180°	56 x 50.4	124.13	1489.63	12000	9.8:1	220	Two	Inject. Bosch	G
1964	275/P	12 V 60°	77 x 58.8	273.81	3285.72	7700	9.7:1	320	One	6-38 DCN	G
1964	365/P	12 V 60°	81 x 71	365.86	4390.35	7300	9:1	380	One	6-42 DCN	G
1965	330/P2	12 V 60°	77 x 71	330.62	3967.44	8200	9.8:1	410	Two	6-40 DCN/2	G
1965	275/P2	12 V 60°	77 x 58.8	273.81	3285.72	8500	9.8:1	350	Two	6-40 DCN/2	G
1965	166/Dino	6 V 65°	77 x 57	265.42	1592.57	9000	11.5:1	175	Two	3-40 DCN/2	G
1965	206/Dino	6 V 65°	86 x 57	331.10	1986.61	8800	11:1	205	Two	3-40 DCN/2	G
1966	330/P3	12 V 60°	77 x 71	330.62	3967.44	8200	10.5:1	420	Two	Inject. Lucas	G
1966	206/Dino	6 V 65°	86 x 57	331.10	1986.60	8800	11:1	240	Two	Inject. Lucas	G
1966	312/F1	12 V 60°	77 x 53.5	249.12	2989.56	10000	11:1	375	Two	Inject. Lucas	G
66/67	330/P4	12 V 60°	77 x 71	330.62	3989.56	8200	10.5:1	450	Two	Inject. Lucas	G
1967	312/F1	12 V 60°	77 x 53.5	249.12	2989.56	10000	11:1	390	Two	Inject. Lucas	G
1967	166/F2	6 V 65°	86 x 45.8	266.04	1596.25	10500	11:1	210	Two	Inject. Lucas	G
1967	350/Can-Am	12 V 60°	79 x 71	348.01	4176.22	8500	11:1	480	Two	Inject. Lucas	G
1968	312/F1	12 V 60°	77 x 53.5	249.12	2989.56	10600	11:1	410	Two	Inject. Lucas	G
1968	246/FL	6 V 65°	90 x 63	400.78	2404.73	8900	11.5:1	285	Two	Inject. Lucas	G
1968	212/E	12 V 180°	65 x 50	165.91	1990.98	11800	11:1	290	Two	Inject. Lucas	G
1968	166/F2	6 V 65°	79.5 x 53.5	266.04	1596.25	10600	11:1	225	Two	Inject. Lucas	G
1968	612/CA	12 V 60°	92 x 78	518.13	6222.16	7000	10.5:1	620	Two	Inject. Lucas	G
1968	Dino 206/GT	6 V 65°	86 x 57	331.10	1986.6	8000	9:1	180	Two	3-40 DCNF/3	G
1969	365/GTB4	12 V 60°	81 x 71	365.86	4390.35	7500	8.8:1	352	Two	6-40 DCN2O	G
1969	246/GT	6 V 65°	92.5 x 60	403.20	2419.20	7500	9:1	195	Two	3-40 DCNF/7	G
1969	246/FL	6 V 65°	90 x 63	400.78	2404.73	8900	11.5:1	300	Two	Inject. Lucas	G
1969	312/F1	12 V 60°	77 x 53.5	249.12	2989.56	11000	11:1	436	Two	Inject. Lucas	G
1969	312/P	12 V 60°	77 x 53.5	249.12	2989.56	9800	11:1	430	Two	Inject. Lucas	G
1969	166/F2	6 V 65°	79.5 x 53.5	266.04	1596.25	11000	11:1	232	Two	Inject. Lucas	G
1969	212/E	12 V 180°	65 x 50	165.91	1990.98	11800	11:1	300	Two	Inject. Lucas	G
1969	612/CA	12 V 60°	92 x 78	518.13	6222.16	7700	10.5:1	640	Two	Inject. Lucas	G
1969	312/B	12 V 180°	78.5 x 51.5	249.25	2991.01	12000	11.5:1	450	Two	Inject. Lucas	G
1969	212/TA	12 V 180°	67.5 x 57.5	205.76	2469.14	12800	11.5:1	350	Two	Inject. Lucas	G
1969	512/S	12 V 60°	87 x 70	416.27	4993.53	8500	11.5:1	550	Two	Inject. Lucas	G
1971	312/B	12 V 180°	80 x 49.6	249.31	2991.80	12600	11.5:1	470	Two	Inject. Lucas	G
1973	365/GT4-BB	12 V 180°	81 x 71	365.86	4390.35	7700	8.8:1	380	Two	4-40 IF3C	G
1974	Dino 308/GT4	8 V 90°	81 x 71	365.86	2926.90	7700	8.8:1	250	Two	4-40 DCNF	G
1974	312/B	12 V 180°	80 x 49.6	249.31	2991.80	12600	11.5:1	495	Two	Inject. Lucas	G
1974	Dino 208/GT4	8 V 90°	66.8 x 71	248.78	1990.26	7700	9:1	180	Two	4-34 DCNF	G

This was followed by an even larger engine, the 512/612/712 series. These, however, were pure racing designs with four overhead camshafts and four-valve heads. The 612 came first, and was the least successful, being intended as a Can-Am racer for the 1968 season. Its 6222 cc (380 cubic inches) and 660 hp at 7700 rpm gave Ferrari high hopes for it, but then he did not know what McLaren and Porsche were up to.

The 512 was a new design—largely the work of Gianni Marelli—for the 1970 season, replacing the P4 as Maranello's main battle tank in long-distance races. Its wins have been few, but it has never been out of contention. This four-cam V-12 with four valves per cylinder and Lucas fuel injection delivers 580 hp at 8500 rpm from 4994 cc (304.6 cubic inches) with a peak torque of 365 lb/ft at 6600 rpm. Cylinder dimensions are moderately oversquare at 87 x 70 mm, like the larger 612 which had both bigger bore and longer stroke at 92 x 78 mm. Only one 6.9-liter (712) long-stroke version of this design seems to have been built.

The final generation GT V-12 has two valves per cylinder canted at an included angle of 92 degrees, and breathes through six Weber carburetors. It delivers 352 hp at 6800 rpm, from 4.4-liters (268 cubic inches). The Daytona engine is now out of production thus marking what is likely to be the end of a glorious era, at least for the U.S. enthusiast.

The first Dino production engine was developed from the powerplant which enabled Ferrari to compete in Formula Two races in 1966. The factory had then tooled up to produce a number of the 206 versions. But the Formula Two effort failed, and the production Dino was equipped with the superior 246 version, detuned to 195 hp at 7600 rpm. It has recently been supplemented by the Dino 308 GT—a 90-degree four-cam V-8 design which puts out 255 hp at 7700 rpm from 2990 cc (182.4 cubic inches).

The specific output of Ferrari's touring car engines has climbed from about 55 hp per liter in 1947 to the 77-85 range in 1974. In the unsupercharged racing engines, specific power output has more than doubled, from 77 hp per liter in 1949 to over 150 today.

In many ways Ferrari engines have always represented the highest degree of sophisticated performance available at any given time. Ferrari has been a pioneer in the relentless pursuit of efficiency—horsepower—performance. From the excellence of the product, one might get the impression that Ferrari has stood in the vanguard of design progress. As we have seen, this has not always been the case, for very good reasons. Ferrari's engines have always tended to be rather understressed instead of designed to break barriers, real or imaginary, of machinery's tolerance to stress. Nowhere has Ferrari's conservative attitude been more apparent than in

1970 Dino 246 engine/transmission.

1974 Dino 308.

ino 308.

1974 BB.   **1974 BB with transfer gear case and transmission.**

his hesitancy to adopt fuel injection, and he still uses carburetors (Weber) for the production car engines.

Incidentally, much of the development work on fuel injection at Maranello after Chiti left was carried out by Michael May, a Swiss engineer who had worked at Daimler-Benz and Porsche. May worked under a rather novel arrangement with Ferrari, an agreement described by Griff Borgeson this way: "He would be paid according to increments in power increase achieved. If there was no increase, there would be no pay." Before returning to Germany in 1965, May had—using Bosch injection—seen a power gain from 165 hp to 220 hp in the 1.5-liter Tipo 158 Formula One engine.

Unquestionably, Ferrari's engine designers have always been very thorough, and they have also had an eye for the simple and for the practical. Although the name Ferrari on an engine has never been a guarantee of racing success, the achievements of the engines have such

grandeur that utter failure can be shrugged off as atypical. And the mind boggles at the amazing number of individual designs—over three hundred probably—that have been produced by Ferrari's small engineering staff over a period of less than thirty years.

Despite the factory's renowned secrecy, the combination that brought success to Ferrari engines is there for us all to see. Ferrari's secret is really no secret at all. When you bring together men with talent, dedication, skill, knowledge and will and give them the tools, the time, the budget and the direction they need, goals become attainable.

Compromises must be made along the way, but the compromises made by Ferrari's engine designers have been made in areas that were not central to Ferrari's idea of what his engines should be. One quality was essential to all, and that quality has never been compromised: Performance. ✦

CHAPTER 6

Bodies Beautiful–
Ferrari
Coachbuilding

by Stan Nowak

Few subjects relating to Ferrari cars are as complex or wide-ranging as that of coachbuilding. A number of coachbuilders have worked for Ferrari over the years and turned out a dazzling if confusing series of models both for racing and touring. With his long background as a Ferrari historian and his years of experience in restoring Ferraris, Stan Nowak is in a unique position to describe the history of Ferrari coachbuilding and comment on the cars themselves. In this chapter, he does just that, beginning with a look at the earliest cars in terms of styling and then moving onwards to the present day. Many of the cars he describes are illustrated here or appear elsewhere in color.

The history of Ferrari coachwork is the story of the best in Italian coachbuilding for the postwar period. In particular, the Ferrari bodies built from 1947 through 1954 were extraordinary in the richness of their variety and the quality of their design and execution. During this early period there were no two Ferraris exactly alike and artisan coachbuilders such as Alfredo Vignale made the most of this opportunity for free expression on a car by car basis. It was a

1950 166 Inter by Touring at Turin Auto Show.

1950 166 Inter by Bertone at Turin Auto Show.

period of flowering and it set the stage for the development of many coachbuilders into large commercial enterprises which today build their own designs on a mass production basis—a feature of the Italian automobile industry which has no parallel elsewhere.

With a long and illustrious history dating back to the Roman chariots and the royal carriages of the various kingdoms which would eventually combine to form Italy, Italian coachbuilders first made an international reputation during the late Twenties and early Thirties of this century, producing outstanding classic and sporting coachwork for such native marques as Alfa Romeo, Fiat, Itala, Isotta Fraschini, Lancia and S.P.A. Most of the finest prewar coachbuilders survived the war years: Carrozzeria Allemano (Turin), Carrozzeria Bertone (Turin), Carrozzeria Stabilimenti Farina (Turin), Carrozzeria Pinin Farina, later Pininfarina (Turin), Carrozzeria Ghia (Turin), Carrozzeria "Superleggera" Touring (Milan) and Carrozzeria Zagato (Milan). All produced bodies for Ferrari during the period under discussion. In addition, there was the brilliant newcomer Carrozzeria Vignale (Turin).

Enzo Ferrari's own background before World War II had brought him into contact with the coachbuilders working for Alfa Romeo. These included Zagato—whose 1750 Alfas of the 1920's have become legend—and Touring, responsible for most of the stunning 8C 2300 and 8C 2900 Alfas of the mid and late 1930's and also the designer-builders of the bodies for Ferrari's two 815's.

Not unexpectedly, the first Ferrari body—on the 125 chassis—was designed and built by Carrozzeria Touring and was derivative of the prewar 815 design, a direction which was to prove a dead end, for it looked bulky and heavy and did not reflect the new silhouette possible with the low line of the Ferrari V-12 engine.

The first cycle-fendered car had a "home built" look about it and was undoubtedly constructed in Ferrari's own Maranello works originally for use as a testing "mule." Late in 1947 a new cycle-fendered version appeared which seemed lower, longer and much better balanced from every angle. It is believed that this car, later sold to G. Besana, was built by Touring. A second identical car was sold to Besana's brother. At least five more cycle-fendered Ferraris were produced which were narrower, rounder, lower, with more inviting grilles, using compound curves more extensively than did the earliest cars. One original example of this style is Briggs Cunningham's car which is on exhibit in his museum.

Ferrari's first coupe was conceived by Allemano and appears to have been evolved from an envelope-bodied roadster. This rather boxy design apparently drew its inspiration from a somewhat prominent rectangular grille. The slab sides were unbroken except for long "eyebrows" over the front wheel wells which proved unpopular until Mercedes-Benz adopted them—to serve as a functional boundary layer

fence—for their 300 SL. Neither the notch-back top nor the plain Kaiser-Frazer-like rear reflected any real visual excitement or originality—and Ferrari never used Allemano again.

The first coupes reflecting a serious industrial design approach were produced by Touring and Stabilimenti Farina.

In 1948 Carrozzeria "Superleggera" Touring came up with a series of designs for Frazer Nash and Bristol of England. One of the designs for the former—a roadster for the Shah of Iran—pointed the way toward Touring's unique and extremely attractive sculpturing exemplified by the 1949 Tipo 166 short (88.6 inch) wheelbase Mille Miglia barchetta roadster and Mille Miglia berlinetta coupe. The subsequent long (98.4 inch) wheelbase berlinettas were ballooned out of the smaller and tighter original—and are not nearly as appealing. In laying out these designs, Touring established a distinctive style which carried over into limited production coupes for the Alfa Romeo 1900 series and several dramatic exercises for the exotic Spanish Pegaso.

The word "Superleggera" means "very light" and signifies a method of construction patented by Carrozzeria Touring, utilizing an inner framework of steel tubing. The outer panels of aluminum were formed over a wooden "buck" (a full-scale model of the car made of wood) and then fastened to the tube frame by rolling the edge of the aluminum panels around the tubing. This resulted in an immensely rigid structure exceedingly strong in relation to its weight.

A wooden buck.

Touring-bodied Ferraris made between 1948 and 1953 are as solid and rattle-free today as when they were new. Although the firm was founded in 1925, the "Superleggera" technique was not invented and used until 1937. But from the beginning, Touring's motto was "weight is the enemy and air resistance is the obstacle."

Touring's design for the Ferrari 166 barchetta was exceedingly attractive in its own right and the car achieved real fame when it won the Le Mans 24 Hour Race in 1949. This barchetta style was also used on the first few 4.1-liter 340 Americas driven by Bill Spear and Jim Kimberly in 1951. Models 195, 212 and 225 also were patterned after the barchetta design—and the last one was built for Henry Ford II in 1953.

At least thirty-six barchettas were produced which include eight on the 4.1-liter 340 America chassis. The equally appealing short wheelbase berlinetta, a coupe version of the barchetta, won the Mille Miglia in 1950 and Briggs Cunningham bought the first one to arrive in the United States. This car, a Tipo 195, with Luigi Chinetti at the wheel, ran in the first Sebring race held in late December, 1950. Eleven were built in all, including two 4.1-liter models.

The long wheelbase berlinettas by Touring were intended for road use only—and one was acquired by Mike Hawthorn for getting back and forth from England to Italy. While less attractive than the racing berlinetta (it had a somewhat bloated look by comparison), it did reflect Ferrari's racing heritage and at least eleven of these were also made.

Incidentally, the forerunner of all these fastback berlinettas was a rather upright notchback coupe, two or three of which were built by Touring during 1949. Each of these cars differed in various body details but they were all fitted with a single carburetor V-12, Cabo center lock wheels and hubcaps.

After 1953 Touring embarked on a new set of designs and produced coachwork for Alfa Romeo, Aston Martin, Maserati, Hudson and the English Sunbeam, but their efforts had begun to lose freshness and individuality. Ferrari never again used Touring. However, the barchetta influence was obvious in other marques, notably the A.C. Ace Bristol, and the style was again echoed by the A.C. Shelby Ford.

The firm of Stabilimenti Farina had been established in 1896 and achieved renown during the Twenties and Thirties with luxury coachwork on the larger Fiat and Lancia chassis. Pinin Farina came from this family business and though he struck out on his own in 1930 he always maintained friendly relations with Stabilimenti Farina. Indeed his company was just around the corner from Farina in the late

1949 166 Inter by Touring.

1940's and it was convenient to move work back and forth depending upon the demands of each firm. For example, although the 1947 design for the Cisitalia coupe had originated with Pinin Farina, he could not produce sufficient cars for Cisitalia's needs, and, on a number of occasions, the wooden buck for the coupe was rolled around the corner so Stabilimenti Farina could produce the same body. These particular cars bore Farina's distinctive badge rather than Pinin Farina's.

In any case, Stabilimenti Farina began building bodies for Ferrari at least two years before Pinin Farina. The first cars included a fastback coupe and a convertible based on the original designs for Cisitalia. At least four of the former and two of the latter were produced during 1949 and 1950. A totally new convertible came through in 1950 with Allemano-like ''eyebrows'' over the front wheels, a raked rectangular flat grille and slab sides. A notch-back coupe arrived two years later, this one with the ''eyebrow'' over the grille. It was Stabilimenti Farina's last body for Ferrari—and the company went out of business later that year.

In 1950 Carrozzeria Ghia began the production of Ferrari coupe bodies of great luxury. This firm had been established in 1915 and in the 1950's was run by Luigi Segre and Mario Boano, a name we will

At Carrozzeria Touring in 1951,

Avv. Gaetoni Ponzoni, left, and Ing. Carlo Felice Bianchi-Anderloni.

1950 166 MM by Touring at Turin Auto Show.

1952 225 by Touring at Monaco.

1952 212 by Ghia at the Paris Salon.

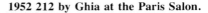

1951 Ghia coupe.

hear again. Segre was the salesman, Boano the designer and together they had a flair for the grand touch. All of their bodies for Ferrari carried overly large and complicated grillework, though in side view they were congenial developments of the trend set by the Cisitalia design. Interiors were the most opulent of all Ferraris of the 1950's; many of the bodies were applied to the large and powerful 342 America chassis. Ghia confined their efforts to Gran Turismo coachwork and did none of the racing or sports-racing bodies.

During the early Fifties Ghia also produced a number of unusual, one-of-a-kind designs for the French firms of Talbot and Delahaye, and Bentley in England. They further specialized in prototype work for Detroit and built a number of special ''dream cars'' for Chrysler which were exhibited at automobile shows throughout the United States in 1952. Although the quality of Ghia's work was beyond question, the firm brought little originality to its Ferrari designs and after 1952 work for Ferrari ceased for three years when a few more Ghia-bodied Ferraris were made. At least twenty-two Ferraris by Ghia were built in all—and all, save two, were coupes.

If Ghia contributed little that was truly original to Ferrari body

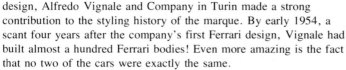
1951 Ghia coupe.

1950 Inter Europa by Vignale.

design, Alfredo Vignale and Company in Turin made a strong contribution to the styling history of the marque. By early 1954, a scant four years after the company's first Ferrari design, Vignale had built almost a hundred Ferrari bodies! Even more amazing is the fact that no two of the cars were exactly the same.

Alfredo Vignale had been brought up in the coachbuilding business as a master metal fabricator and by 1947 was foreman at Carrozzeria Stabilimenti Farina. He then joined Cisitalia where he was responsible for producing three special aerodynamic coupes with sharply vee'd windshield glass and dramatically large fins rising from the rear fenders. On each side of the front fenders were two portholes which General Motors stylists noted and quickly adopted for the 1949 Buick. Piero Dusio, the owner of Cisitalia, was so pleased with Vignale's work that he gave the designer a bonus of one hundred thousand lire. At about the same time, Vignale had built a special streamlined body on a Fiat 500 Topolino chassis which was written up admirably in the press but wrongly attributed to another coachbuilder. The attendant correction and apology spread Vignale's reputation even further. In 1948 he started his own carrozzeria and joined the group of leading coachbuilders of Turin.

For the first few years he specialized in creating coachwork for Lancia Aprilia, Fiat 1100 and 1500 chassis. His work for Ferrari began in late 1949 with a fastback coupe on a Tipo 166 Inter long wheelbase chassis. The overall silhouette was not dissimilar to designs of other leading coachbuilders but the detailing was especially distinctive and captivating. In general, Vignale's early bodies for Ferrari achieved their special appeal through the use of a long curve beginning with the front fenders and sweeping back through the doors and into the rear fenders. Grillework and bumpers (when used) were carefully thought out and executed, as was the special exterior and interior hardware fabricated by Vignale. Parking lights and taillights for Vignale Ferraris were never the over-the-counter production items used by most coachbuilders; Vignale invariably built his own, and these details were integrated parts of the complete design. Another unique Vignale feature was his tasteful use of chrome-plated or polished aluminum beading to delineate a color separation line on two tones, usually between the top and sides of the car.

1951 212 Inter by Vignale at the Paris salon.

1951 166 MM by Vignale.

Making body panels at Farina.

Most of Vignale's designs during 1950 and 1951 were the work of Giovanni Michelotti, a prolific and talented freelance automotive stylist who also worked with Ghia-Aigle in Switzerland (builders of at least two of his designs on Ferrari chassis during 1951). Michelotti and Vignale developed an extraordinary rapport and their cars produced for Ferrari were perhaps the most exciting built during this period.

Unquestionably, Alfredo Vignale was a genius in being able to translate Michelotti's drawings into metal without the use of a full-scale buck, proceeding from full-scale drawings and sketches directly into aluminum. This method was unique to Vignale and resulted in cars which were slightly asymmetrical, just as the left side of the human face is not the same as the right side. Some experts have concluded that this makes the cars even more visually interesting and endows them with more character than the ordinary run of vehicle. The differences were more than purely superficial; measuring a Vignale-bodied Tipo 166 Mille Miglia coupe some time back showed a left hand door two inches longer than the right hand door.

Vignale worked exclusively in aluminum during this period, using the time-proven Italian method of forming metal which began with the shaping of the first Italian suit of armor, to wit: the flat sheet is first rough-shaped on a tree-stump (sic) using a wooden mallet; it is then smoothed out, using a flat hammer on top of a sandbag, the final shaping and finishing done with a flat hammer on an anvil. From time to time during this process, the piece is held up to the car to check for fit. The finished aluminum panel is finally positioned in place by folding the metal around the steel framework, and held fast by rivets and screws. The body then receives a liberal coat of filler known as ''gumite'' to the Italians and ''Bondo'' to Americans. This is worked to a smooth finish before the paint process begins.

As Vignale's cooperation with Ferrari continued into 1951, his output shifted in emphasis from the long wheelbase Inter chassis to the short wheelbase competition models, the 212 Export and the new 225. Vignale seemed to specialize also in bodying future Mille Miglia winning chassis: both the 1951 340 America coupe (driven by Villoresi) and the 1952 250 S coupe (Bracco driving) were outright

ca 1951 195 Inter by Vignale.

winners! This certainly helped to increase Vignale's popularity and by 1952 his firm was building an ever larger proportion of competition cars for Ferrari.

That year too four special cars were designed by Michelotti and built by Vignale for the Mexican Road Race (more correctly known as the Carrera Panamericana). Three were coupes with particularly "wild" styling which worked well for racing machines but became gross and overdecorated when considered as luxury touring cars. The fourth was a roadster which never ran in Mexico.

Vignale's most admirable designs were for the short wheelbase competition roadsters and coupes made during 1951 and 1952. These cars had a concise, clean, taut look—curvaceous yet restrained—that Vignale was never able to perpetuate. This purity evolved into the more complex lines of the Mexico coupes. Vignale styling took a new direction in late 1952 and 1953 with a run of 166 Mille Miglia, 250 Mille Miglia and 340 Mille Miglia roadsters typified by Giannino Marzotto's Mille Miglia winning 1953 roadster. This design resurrected the Cisitalia's use of portholes. The front of the car was unique in forsaking the use of two headlights as beginning reference points for the overall design; instead, the envelope was devised for the most pleasing curves and the headlights were buried below the

166 by Farina at 1950 Paris Salon.

1952 225 by Vignale at Monaco.

1952 Vignale coupe.

1952 Mexico.

166 or 250 MM by Vignale.

340 MM by Vignale.

139

250 Europa by Vignale at 1954 Geneva Show.

340 coupe by Vignale at 1954 Geneva Show.

bodyline in toward the egg-crate grille. The backs of the rear fenders were cut out to encourage the exiting of hot air from the wheels and brakes. A gutsy, aggressive design, to be sure, but it failed in being based on a doughy overall form.

By the summer of 1953, Vignale had completed the last body for the 250 Mille Miglia series and had been phased out of Ferrari's coachbuilding program. Phased in was Pinin Farina. Vignale did build two more special bodies for Ferrari in 1954, a convertible and a coupe, both with American style wraparound windshields, but this was the end of any further cooperation with Ferrari.

The first Pinin Farina-bodied Ferrari, a dignified, tidy convertible on a 212 Inter chassis, was shown at the Paris Salon late in 1952. This was followed early the next year by a variety of convertibles and rather dowdy coupes on the new 250 Europa and 342 America chassis, exhibiting a conservative approach that was in stark contrast to the more flamboyant Vignale designs. At their worst, these early Pinin Farina efforts were lumpy and tentative and, at best, only satisfactory. Slowly, Pinin Farina's designs improved, and the first series 250 Europa coupe introduced in 1953 reflects this. About twenty of these cars were built on the 110 inch chassis. At the beginning of 1954 the second series Europa GT was introduced on the 102.3 inch chassis. The eight inches were taken out of the cowl just forward of the windshield but the cars themselves were otherwise almost identical to

1952 212 Inter by Pininfarina.

1952 212 Inter by Pininfarina at Paris Salon.

Circa 1952 212 or 250 Europa by Pininfarina.

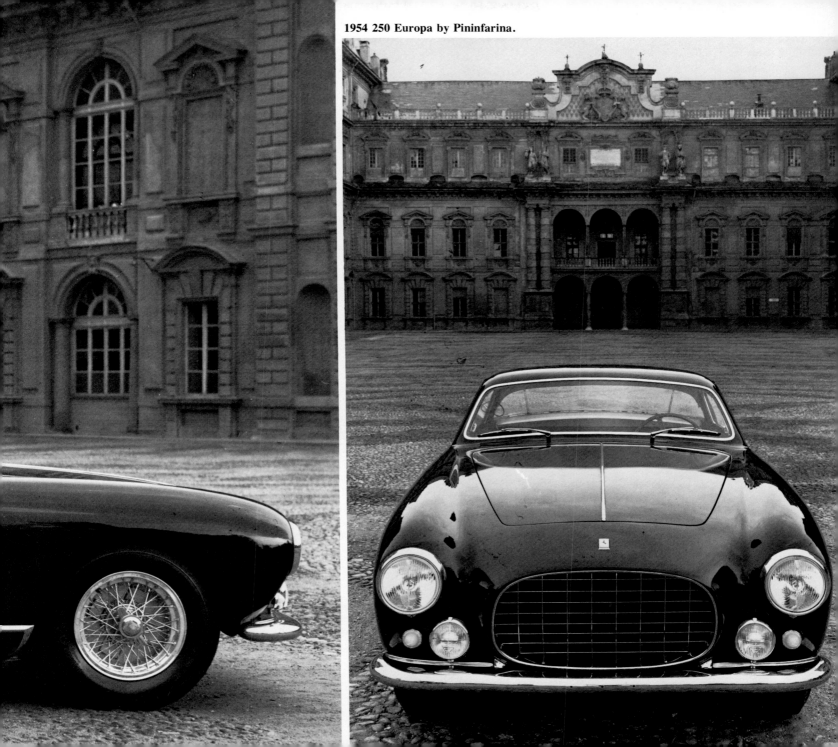

1954 250 Europa by Pininfarina.

1953 250 MM by Pininfarina.

1953 375 MM by Pininfarina.

1953 375 MM by Pininfarina.

those of the initial series. Second series Europas were the first Ferraris to use the GT suffix after the serial number; the first series Europa numbers bear the suffix EU. The Europa GT was Ferrari's initial experiment in marketing a semi-production touring car and continued to be built through the early months of 1956.

From late 1952 and throughout 1953, Pinin Farina was also producing a superbly sculptured fastback coupe body on the 250 Mille Miglia competition chassis. Like the early Vignale designs, this coupe was a deceptively simple, uncluttered entity which depended on subtle shaping of the metal to achieve its unified look. Of the thirty-two 250 Mille Miglias built, seventeen were Pinin Farina coupes, fourteen were Vignale roadsters and one was a Vignale coupe.

During 1953 Pinin Farina also designed and built a more limited number of all-aluminum competition bodies on the 4.5 liter 375 Mille Miglia and two-liter, four-cylinder Mondial chassis. The 375 MM bodies were closed coupes with a shark-mouth grille opening and a more aggressive overall look than the 250 MM coupe. The first Mondial bodies were roadsters and were, in effect, Pinin Farina's idea of what a barchetta should look like. The result was extraordinarily effective, establishing a new standard for the ultimate in form, delineating the function of a two-place sports-racing car. The design was so good that Pinin Farina used it again in 1954, on a slightly

1954 375 by Pininfarina at the Geneva Show.

d Bergman's 1954 375 America.

larger scale, to clothe the 375 and 375 Plus Mille Miglia chassis. A number of 375 Plus Pinin Farina roadsters, usually called spyders when reference is to a racing car, were entered for the 1954 Le Mans 24 Hour Race and one of them, driven by Gonzalez, became the overall winner.

Two special order Pinin Farina Ferraris of 1954 and 1955 are also worthy of note, the first commissioned by Roberto Rossellini (for Ingrid Bergman), the second by ex-King Leopold of Belgium. The Rossellini car was a most unusual and pleasing coupe on a 375 Mille Miglia chassis. The design incorporated a number of innovative features: there were no visible headlights (they were the pop-up type so popular on many sports cars twenty years later), the sides of the front fenders included concave airfoil shapes just behind the front wheels (this feature was adopted by Chevrolet for their 1956 Corvette), and at the rear, buttresses flowed aft from either side of the rear window (this styling device became almost a Pinin Farina signature on various designs over the next twenty years). The ex-King Leopold car was a convertible on a 375 America chassis—and it was, as they say, a "looker." Its swooping elegance has never been bettered and even the long wheelbase California spyder produced three years later did not achieve the startling and long-lasting beauty of this one-of-a-kind creation.

Leopold's 1953 342 America.

1955 250 GT by Pininfarina.

To all outward appearances, Ferrari had now decided to use Pinin Farina exclusively for all coachwork on Ferrari cars. But this was deceiving. In fact, Enzo Ferrari's son Dino had begun to experiment with competition sports car coachwork made by the small shop of Franco Scaglietti in Modena. A one-off spyder body with lines suggested by Dino on an early Tipo 166 Mille Miglia chassis had been the first result of this collaboration and by mid-1954, several 500 Mondials and the rare 250 V-12 Monzas had received the Dino-inspired Scaglietti spyder coachwork as well. This led to the four-cylinder Monza spyders of 1955 and a relationship between Ferrari and Scaglietti which survives to the present day.

At Pinin Farina, meanwhile, work proceeded on a design for a production 250 GT coupe. By 1955 several prototypes had been built based on the later 250 Europa GT's but with a flatter, elliptical grille. Pinin Farina's factory was now large enough to handle production of these cars on an assembly line basis if a minimum of five hundred cars were ordered. Ferrari, however, did not want to commit himself to such a large order so the actual production was given instead to Boano who had by now left Ghia. Luigi Segre, Ghia's manager, preferred to specialize in the construction of prototype cars on an individual basis; Boano wanted to produce cars in quantity. It is believed that Ferrari ordered these first-production 250 GT's in batches of one hundred.

147

The design was Pinin Farina's with only very slight beltline simplification—and no coachbuilder's badge. The first of these cars was completed early in 1956 and, at the same time, Boano designed and produced a prototype 250 GT convertible with a bold new front, handsome sculptured side view and faddish curved fins sweeping outward on either side of the rear. The design was not accepted, however—and the following year Pinin Farina began producing a series of 250 GT convertibles which culminated in the introduction of the 250 GT Pinin Farina cabriolet in late 1958 as part of the 1959 range of models.

From Pinin Farina's 375 MM berlinetta of 1955 came the first of the so-called Tour de France 250 GT coupes. Production of this small series was entrusted to Scaglietti. This model was intended for competition in such events as the Mille Miglia, Tour de France and 1000 Kilometers of the Nürburgring. The body was built entirely of aluminum, with every effort made to eliminate heavy fittings. With a curb weight of only 2520 pounds, it was at least five hundred pounds lighter than the standard Boano coupe—and its three-liter engine was developed to produce 260 hp at 7000 rpm. The car was an immediate success and in 1956 won the Tour de France outright, driven by de Portago—and another won its class in the Mille Miglia with Gendebien at the wheel.

Scaglietti could not produce enough of these new berlinettas to meet the demand so five chassis were given to Carrozzeria Zagato (whose last body for Ferrari had been the one unusual Panoramica coupe in 1950). No two Zagato bodies were the same but all were notch-back coupes built entirely of aluminum and benefiting greatly from Zagato's long years of experience in devising lightweight competition cars. Two of the cars utilized the "double-bubble" roof design which had attracted much attention when applied to Zagato's delightful design for the tiny Abarth 750 cc coupe. This was the end of Ferrari's official cooperation with Zagato. (In 1971 Luigi Chinetti had Zagato build a unique convertible body on a 250 GT chassis and in 1975 two more bodies were commissioned by Chinetti on older chassis, one a coupe with a removable roof, the other a convertible.)

The design of the 250 GT Tour de France berlinettas, originally based on Pinin Farina's 375 MM berlinetta, evolved over the years. The early cars made in 1955 and 1956 are distinguished by the combination of a wraparound rear window and "frenched" headlights sunken into the very front of the fenders with no visible outer rim. The 1957 version had a more pronounced, flatter, elliptical grille opening, the headlight treatment remaining unchanged. The rear window no longer wrapped around and the metal area behind the door windows was now decorated with thirteen or fourteen louvers diminishing in size going aft. Neither of the first two styles were

148

1956 250 GT by Pininfarina.

1957 250 GT by Pininfarina.

1961 250 GT—prototype GTO by Pininfarina.

equipped with full bumpers, small vertical bumperettes sufficing.

Although these cars were built by Scaglietti, the designs were influenced by Pinin Farina and the result was really a collaborative effort. This is borne out by the fact that neither the Scaglietti nor the Pininfarina (as the coachbuilder had legally changed the spelling of his name about now) badge was ever affixed to any of the Tour de France models. For 1958 a further redesign was carried out which included raising and moving back the headlights and fairing them into the fender line with plastic covers. The multi-louver panel was replaced with three rectangular vents and the rear fenders were carried higher, ending in fender tip taillights. Simple full-width bumpers of aluminum were added, apparently to give the appearance of a normal road model.

The 1959 Tour de France was the same as the '58 except for upright exposed headlights and one vent in place of three behind the door windows. On this same 102.3 inch "Long Wheelbase" chassis, Scaglietti began building the new 250 GT Spyder California which had been designed by Pinin Farina for Luigi Chinetti. Production of this new model, which might have been called a Tour de France roadster, began in the winter of 1958 and continued through 1959. The lines were quite similar to the 1958 Tour de France berlinetta but with a flatter windshield, giving less wraparound at the sides. Unlike the all-aluminum berlinettas, almost all of the Californias were made of steel with aluminum hood and trunk lid only. However, in mid-1959, at least one LWB California was made entirely of aluminum and had the upright exposed headlights of the '59 Tour de France.

By the spring of 1958, Scaglietti was building an entirely new berlinetta on the LWB chassis. Pinin Farina had designed this body and built at least one of the six that were made. The rest were by Scaglietti who was now well established as the specialist in producing Ferrari's dual-purpose berlinetta cars. The new design incorporated fewer straight lines than the Tour de France and its most distinctive features were the angular quarter windows behind the door glass and a shark-like nose, raked under, with the traditional egg-crate grille buried deep inside. This has become known as the "interim" berlinetta as so few were made and it was immediately superceded, later in 1959, by a new short wheelbase (94.5 inch) model.

From 1956 through 1959 Scaglietti continued to build and design, with the collaboration of Pinin Farina and Ferrari, a much smaller number of two-seater racing cars. In 1956 these included the 860 Monza, whose styling was directly derived from the 750 Monza, the 410 Sport and the new 500 TR four-cylinder Testa Rossa. The latter was a two-liter sports-racing car intended for customer racing in the international two-liter class. The chassis was much simpler than the earlier two-liter Mondial, with a live rear axle and the gearbox mated to the engine rather than the expensive transaxle/de Dion layout. This

1957 250 Testa Rossa by Scagli

1961 250 TRI by Fantuzzi

was a less voluptuous design than the Monza with a thinner, lower front grille.

Nineteen fifty-seven saw the introduction of two of the finest designs ever produced by Scaglietti: the "low-bodied" 500 TRC roadster and the strikingly original pontoon-fendered V-12 Testa Rossa. Both models have been cited by most Ferrari connoisseurs as the most pleasing produced for a two-place racing car. Both carried the Scaglietti badge and are a tribute to his ability as a designer and fabricator. Without a doubt, these cars were the ultimate development of open sports-racing car styling for the front-engined configuration.

The 500 TRC was three inches lower than the 500 TR and conformed to the 1957 Appendix C regulations for sports-racing cars. At the very front of the car was the extremely wide, slim grille opening edged in polished aluminum and leading to a much recessed egg-crate grille. The lights were buried in the fender forms and covered with plastic to match the fender line. The side view was particularly attractive with the well-rounded front fenders sweeping down to the low-cut doors and up again to cover the rear wheels in another sweep of curved aluminum. Overall, it is one of the most cohesive and finely balanced designs ever conceived.

The somewhat startling V-12 250 Testa Rossa was formally announced at the factory press showing in November, 1957, although prototypes had been raced at Nürburgring and Le Mans earlier in the year. Actually this new V-12 model replaced the 500 TRC which was discontinued at the end of 1957. Looking head on, the car was unique in treating the front of the two fenders and the air intake grille unit as three separate elements; in side view, the most notable feature was the carrying of the front fender line around the tire and straight back from the top of the tire into the bodywork just forward to the door. The rear of the car was similar to the 500 TRC but had the fenders peaking aft to small taillights. This design was made available to Ferrari's racing customers and about eighteen were produced in total through early 1959. Over the next three years the 250 TR was further developed in a series of new designs that were laid out by Pininfarina and built by Fantuzzi whose modest shop had achieved fame in Modena for building almost all of the Maserati racing cars of the early 1950's.

At the beginning, these bodies resembled the old 500 TRC design but with a large plexiglass intake over the carburetors. In 1961 a new

56 Superfast I by Pininfarina.

Sweden's Prince Bertil owned this 1958 250 GT by Pininfarina

body appeared with an extended nostril nose, a high rear body and a spoiler across the rear which has since become an important aerodynamic device on racing cars and a popular styling device on road cars. Californian Richie Ginther, who was Ferrari's chief test driver at the time, is credited with the invention of this device which was originally intended to block exhaust gases from coming forward to the cockpit when decelerating but was also found to improve the car's stability at high speed. The spoiler was tested in Italy and first used at Sebring in March, 1961 when Phil Hill won outright in a 250 TRI-61.

A final design was developed for 1962. Known as the 330 TRI, the body was created and built by Fantuzzi and is unusual in utilizing angular squared-off lines that are not entirely pleasing and, in fact, proved to be a dead end. It also featured a ''basket handle'' airfoil-cum-rollbar just behind the cockpit which found favor with Ferrari and was to be seen again on the rear-engined sports-racing cars. The car itself was a great success, having won the 1962 Le Mans 24 Hour Race driven by Phil Hill and Oliver Gendebien. This was the last of the great front-engined prototype sports-racing cars.

Pinin Farina's first major styling prototype for Ferrari was the Superfast I introduced in 1956. It was built on a special 410 Superamerica chassis using a twin ignition 410 Sport racing engine. The design incorporated a large number of new and unusual details many of which were adopted over the next few years on most of the

production cars and even some of the racing cars. The design of the Superfast I's front—its headlight treatment and grill opening— exerted the strongest influence on subsequent Ferraris. The car's most startling features, its pillarless windshield and fins, were not used again.

In 1958, when Pinin Farina legally changed his name, he also changed his company's to Pininfarina and left its running largely to his son Sergio Pininfarina and his son-in-law Renzo Carli. There followed the launching of three new production Ferraris for 1959: the 250 GT convertible, the 250 GT coupe and the 250 GT short wheelbase berlinetta. In addition, the company continued to make a few special order 410 Superamerica coupes and convertibles. These limited production cars had been introduced at the Paris salon in late 1955. The Pininfarina-designed and built body of the Superamerica was a luxuriously-detailed, slightly larger version of the Europa GT's styling though in many cases more gaudy.

The production 250 GT convertible was directly based on a very limited number of convertibles Pininfarina had been making since 1956. The styling was understated and did not reflect Ferrari's racing heritage. This model was entirely built by Pininfarina and featured full bumpers front and rear, exposed headlights, vent windows in the doors and pronounced wraparound of the sides of the windshield. An optional extra was a Pininfarina-designed removable hard-top with a large rear window. With only very minor trim changes, this model

Circa 1963 250 GT by Pininfarina.

remained the same through 1962 when production ceased.

After building three prototypes, Pininfarina began production of the new 250 GT notch-back coupe in late 1958 as a 1959 model. Again, the design was markedly restrained and showed a subdued grace which was to be the hallmark of all Ferraris intended only for touring. Certainly, these designs were a disappointment to the true Ferrari aficionados who saw the marque only in a competition context. As a result, this model has perhaps been underrated and overlooked. It is, nevertheless, one of Pininfarina's most elegant and classic Ferraris and can be judged today as an outstanding design.

At the end of 1959, the third new Pininfarina design was ready for production. It was the 250 GT short wheelbase berlinetta. This was developed from the "interim" long wheelbase model (the end of the Tour de France line) and the first prototype appeared in mid-1959. By September a production version was shown at the Paris Salon. The bodies for these cars were made by Scaglietti on a production line basis. Unlike the other two models the SWB berlinetta was intended as a dual-purpose car and a number of all-aluminum ones were built for competition. However, the steel-bodied cars were also raced, as the aluminum version was only about a hundred pounds lighter. The Pininfarina design has the taut, muscular look of a wild animal ready to jump. This appearance of metal in tension conveys the purpose of the car and it is a design which has stood the test of time. Despite the

competition success of the SWB berlinetta, production of this model was phased out by mid-1962.

By the end of the 1961 season, the demands of GT class racing indicated the need for a lighter, lower, more powerful car and Giotto Bizzarrini, a bright young Ferrari engineer, was put in charge of developing one. This new car was to be called the GTO.

It evolved from improvements to the SWB berlinettas, a number of which were equipped with the six dual-throat Weber carburetors which had been developed for the V-12 Testa Rossa and were also used on the new GTO.

Pininfarina built several prototype GTO's based on the Superfast II design which was later used for the semi-production 400 SA coupes. This design, however, did not find favor and engineers Bizzarrini and Carlo Chiti, Ferrari's chief engineer, suggested new lines to Scaglietti who did the actual panel beating. The new car was first shown at the Ferrari press gathering in February, 1962 and the lithe design took full advantage of the low line made possible by the dry sump engine. All the GTO's were made of light-gauge aluminum and used six dual-throat Weber carburetors, special heads and cams to extract 300 hp from the three-liter V-12 engine.

The overall design was created to "do a job," but the aesthetic result was the most appealing Ferrari berlinetta yet made. The front overhang was greater than previous models to allow for a very low,

153

1956 410 Superamerica by Pininfarina.

1957 410 Superamerica by Pininfarina.

1958 250 GT special by Pininfarina.

1960 Ferrari 400 Superamerica by Pininfarina.

Circa 1958 250 GT special by Pininfarina.

elliptical radiator air intake at the front of a long, sloping hood. The headlights were positioned low in the fenders behind fully faired plastic covers. Smooth clean lines were used throughout with no tricks except for diagonal rectangular openings on the sides just ahead of the doors and D-shaped scoops behind the rear wheels. The roof sloped gently back behind the well-raked windshield to a truncated rear topped by a full-width spoiler. At least forty of these cars were built during 1962, 1963 and 1964. The three GTO's built in 1964 were fitted with bodies which used a 250 LM coupe-style, tunnel-back top and two of the fast-back GTO's were rebodied in this fashion. This latter body style usually is referred to as the Series II.

Unquestionably, the GTO was one of the greatest GT cars ever built. And let there be no argument about the GTO being a GT automobile in the truest sense. Anyone who has owned or driven one of these cars in recent years has come to realize how remarkably tractable the GTO really is. It is an honest dual-purpose car with an overall performance that is hard to beat even today. The GTO's worth is accurately reflected in its present status as a collector's item; it is valued higher than any other older sports-racing Ferrari.

The 1960 season saw the introduction of a short wheelbase California roadster along the lines of the previous long wheelbase version. It is not easy to tell the difference between these two models without measuring the wheelbase. If the chassis/engine number is below 1700GT it is a LWB, above 1700GT it is a SWB. However, if the two versions are seen side by side, the lines of the long wheelbase are visually more pleasing. The new model was also designed by Pininfarina and made by Scaglietti in Modena.

At the end of 1960 Pininfarina introduced two more new cars, the 250 GTE 2+2 and a prototype Superfast II for Enzo Ferrari's personal

156

250 GT E 2+2 by Pininfarina.

1963 250 GT Lusso by Pininfarina.

1960 250 GT 2+2 by Pininfarina.

use. The former was Ferrari's first serious effort to market a luxury "family car" with at least the possibility of holding four people. The design had been carefully studied over several years—including extensive wind tunnel tests which resulted in the car's 150 mph top speed—and three final working prototypes were built before an ultimate selection was made. The clean straight lines with a minimum of decoration are representative of the dignity of all the Pininfarina creations. The forward thrust of the headlights and taillight assemblies give the design a dynamic balance that is most impressive for a car of its type and size. The large amount of glass area was also exceptional and the overlapping of the rear window and rear quarter vents contributed to the car's dynamic qualities. Unlike previous production cars, a great effort was made to provide the 2+2 with a truly well-designed and luxurious interior whose specification included individually reclining bucket seats. For the first time air conditioning was offered as a factory-installed option. With only very minor trim changes the 2+2 remained the same through 1963.

In February, 1961 Ferrari introduced their first rear-engined sports-racing V-6-engined Dino car, the 246 SP. The body design was the result of wind tunnel tests initiated by engineer Chiti and most easily identified by the attractive nostril nose and full wraparound windshield extending into a high rear body punctuated by a truncated tail and topped by a full-width spoiler. Chiti's ideas were carried into metal by Fantuzzi.

The 246 SP had been developed from a series of V-6 front-engined sports-racing cars beginning with the Dino 206 S and Dino 296 S in 1958. These were followed by the Dino 196 S in 1959 and the Dino 246 S in 1960. In all cases, the coachwork was almost identical to that used on the factory-entered competition V-12 Testa Rossa of that year.

157

In February, 1962 Ferrari introduced, along with the previously discussed GTO, three new rear-engined cars: the 196 SP, the 286 SP and the first V-8-engined sports-racing car, the 248 SP. These three types bore identical Chiti-developed coachwork. This design had been evolved from the 246 SP and again used the nostril nose, but this time it was distinguished with a horizontal air intake slot just above the nose. The full wraparound windshield was used but was much lower in order to match the new lower silhouette of the rear. Although they carried no coachbuilder's badge, these bodies were constructed by Fantuzzi. The cars were raced in Italy and the United States for the next two years and no further Dino V-6 sports-racing car was introduced until 1965.

Over the years Ferrari production had risen steadily and in 1962 a large addition to the Maranello factory was constructed to accommodate additional production lines. At the end of the year the luscious new 250 GT Lusso berlinetta was introduced. This was a GTO-inspired, Pininfarina-designed, Scaglietti-built berlinetta whose lines brought sighs from critics throughout the world. Indeed, it was an inspired design as clean as Pininfarina's previous studies but using more zaftig, smaller-radius curves which attracted a whole new clientele. The tail was truncated as on the GTO and the top surface swept upward as a vestigial spoiler.

Despite its sensual lines, the Lusso was no replacement for the SWB berlinetta. The serious Gran Turismo competition was in the province of the GTO and the Lusso was more in the performance category of a two-passenger version of the 2 + 2. The only exceptions were four all-aluminum sports-racing berlinettas utilizing the Lusso lines with the addition of a GTO nose. Of these special Scaglietti-built factory cars, three were 330 LMB's, and one was a 250 GTO.

Although exceptionally well designed, the Lusso was a disappointment to the serious Ferrari enthusiast for it was not a true dual-purpose car as the Tour de France and SWB berlinetta models had been. Within only a year and a half, the Lusso was being phased out and by August, 1964 the last few Lussos were delivered fitted with the new 3.3-liter 275 series engine.

In 1961 Pininfarina's super luxury 400 Superamerica was introduced, based on the Superfast II design that had been specially made for Enzo Ferrari. This very limited production model was made

250 Le Mans by Pininfarina.

to special order in Pininfarina's own shops and production was restricted to about two per month at a price which was almost double that of the production SWB berlinetta. The lines of this model established a new direction for Pininfarina and parts of this design later found their way into some of the production cars. The basic 400 SA design lasted through 1964 although it was slightly changed at the end of 1962 when built on a longer wheelbase chassis.

Toward the end of 1963 the 250 GT SWB California roadster was discontinued and Ferrari was without an open car in its production lineup until 1965. At the same time the 250 GT 2+2 was being delivered with a new more powerful four-liter 330 GT engine. A few months later a totally new roomier 330 GT 2+2 made its debut. Its most obvious feature was four headlights. The rest was a bland restatement of the previous design, although it was built on a longer wheelbase and the interior was larger in every dimension. The four headlights drew instant criticism from every quarter and by mid-1965 they had been quietly dropped. The design remained otherwise the same and continued in production through 1967.

For the 1964 racing season Ferrari planned to enter cars in both the prototype and GT categories. The prototype cars were the 275 P and 330 P and were smoothed out versions of the 250 P. The "basket handle" rollbar was still there but with a side window connecting it to the windshield. As before, the body was the work of Fantuzzi's shop in Modena. For the GT category Ferrari introduced a replacement for the GTO in the form of a rear-engined coupe: the 250 LM. The design was the work of Pininfarina's studio although all but one or two special show models were made by Scaglietti. The 250 LM was basically a 250 P with a roof. A vertical rear window was installed in what had been the "basket handle" rollbar structure behind the cockpit. Full doors were used with plexiglass sliding windows. The first car incorporated an airfoil across the car at the rear of the very short roof. This did not appear on the later versions and the entire top was lengthened somewhat to improve the aerodynamics of the car as well as its appearance. This proved to be the definitive shape of this model despite several experiments with a large wraparound rear window extending back from the top, as well as a similar fastback configuration.

Production continued through 1965 and the 250 LM won the 1965

275 GTB by Pininfarina.

1967 Dino 206 GT by Pininfarina.

Le Mans 24 Hour Race. As Ferrari was not able to get it homologated for GT racing, the GTO was the last true GT model built by Ferrari. The 250 LM's ran as prototype cars. Because of his disappointment over the 250 LM, Ferrari abandoned any further efforts at factory-sponsored GT racing except for one special lightweight 275 GTB which finished third at Le Mans in 1965, winning the GT class.

By September, 1964 Ferrari had phased out the 250 GT Lusso berlinetta and had begun to deliver a completely new 275 GTB berlinetta which broke new ground in Ferrari production carchassis design by including independent rear suspension and a transaxle in its

specification. Another innovation for Ferrari was the use of knock-off magnesium wheels in place of the traditional glistening wires. Designed by Pininfarina and built by Scaglietti, the 275 GTB made a strong design statement and was more in the spirit of the sports-racing GTO than the more effete Lusso. Even the four slanted vents on the sides of the front fenders echoed the GTO design. The racing look was further strengthened by the use of a large well-raked windshield similar to that used on the Series II GTO and the 250 LM. For all practical purposes GT racing was left to Ferrari's private customers and the GTB could be ordered with a variety of competition-oriented

330 P3 by Cigario.

options including an all-aluminum body with a long nose, dry sump
lubrication, six carburetors with higher compression heads, and large
outside filler cap. It was the first dual-purpose Ferrari since the SWB
berlinetta was discontinued in 1961.

Other completely new catalog production models were introduced
in 1965. These were the 275 GTS convertible, the 330 GTC
coupe—both by Pininfarina—and the more limited production 500
Superfast coupe. It was a most ambitous program as all of these
models and the 275 GTB were entirely new.

The 275 GTS convertible was a luxury concept but its design was

soft, bland and undistinguished, having evolved directly from the 330
GT 2+2. The 330 GTC coupe was a different proposition. The
attractive character of its design derived from its long low
Superfast-type nose, indented belt line, and its angular side glass areas
contrasting the rounded front and rear window shapes. A splendidly
fitted leather interior with reclining seats completed the appealing
design. In effect, the 330 GTC took the place of the Lusso as the
epitome of a powerful two-passenger high-speed deluxe touring car.

The 500 Superfast was an entirely new replacement for the 400
Superamerica and production at Pininfarina's plant continued on the

161

1968 250 P5 by Pininfarina.

basis of only one or two individually made cars each month. This design—a cleaned-up version of the 400 Superamerica with a truncated tail—was replaced in 1967 with the 365 California convertible. It had the same clean lines as the 500 Superfast but with a scooped-out horizontal depression in the top of each door leading to the rear with a metal spear running across the top. A similar motif was used later in the production Dino 206 GT coupe where it served as a functional air intake for brake cooling. On the California—of which only a few were made—it was merely a styling device to cover the door latch.

In December of 1964 Ferrari had announced the Fiat-Ferrari Dino 166 project which included the development of a Dino V-6 engine to be the basis of a new Fiat Dino sports car with the engines to be produced by Fiat according to Ferrari designs. In April, 1965 a new 166 P sports-racing coupe appeared and the Dino badge on the nose was the first step in establishing the Dino as a separate marque. This miniscule racing coupe was delightfully formed with curves on almost every surface. The cockpit was surrounded in curved plexiglass, giving a light and airy feeling to the roof. The design was based on a scaled-down version of the powerful P2 prototype car introduced at the same time and both were the result of design work carried out by Ferrari engineers using wind tunnel studies to arrive at solutions that worked best on the track. The actual bodies were made in the small shop of Piero Drogo who occupied the front of the same building that housed Fantuzzi.

Ferrari continued to develop his prototype racing cars and in 1966 he introduced the eye-catching new 330 P3 spyder at Sebring. A berlinetta version was given its debut a few months later and, in addition, most of the old P2's were rebodied with the new P3 coachwork. Thereafter, the latter were known as P2/3. This dramatic design was developed by the Ferrari engineers including Mauro Forghieri who was in charge of engineering development of all the factory racing cars. The aluminum body was made by Cigario, another small Modenese shop. As sports car prototype racing became more competitive, the designs of the cars became more and more dictated by pragmatic considerations (e.g., how to get air into and away from the radiator) and the ''state of the art.'' Many of the styling decisions were based on wind tunnel tests and changes were quite often made as the result of testing at the track and during actual races. As the designs became increasingly specialized, the ''stylist'' had less and less to do with influencing the final shape.

In 1967 the P3 was replaced by the P4 which appeared first at Daytona in February. The P4 design was only slightly altered from the P3 and again Cigario was responsible for making the bodies. The P4 was the end of the line for the well-rounded ''soft'' curved body designs. Ferrari dropped out of sports car racing during 1968 (except

163

512 S by Pininfarina.

1970 PF Modulo by Pininfarina.

1969 365 GTS/4 by Pininfarina.

1969 365 GTB/4 by Pininfarina.

for one outing with the 612 Can-Am spyder) and returned in 1969 with a much subdued 312 P spyder and berlinetta. Similar in form were the 612 Can-Am and the first twelve-cylinder opposed "boxer" engined sports racer, the 212 E spyder used for hill climbs. In December of 1969 the 512 S appeared and this berlinetta ushered in a new era of squared-off designs based ninety-five percent on wind tunnel tests and other engineering considerations. These were followed in 1970 by the 312 PB (B for Boxer) spyders and are of interest as styling exercises only as they provided feedback to the designers of the production cars.

At Pininfarina racing car design influence was manifest in a number of show cars built and displayed from 1967 through 1970. The first was a berlinetta Dino Special with wing spoilers at both ends and a huge rounded windshield. More subtle and impressive was a 1968 showpiece called the 250 P5. Using basically soft lines derived from the racing P4, Pininfarina adroitly integrated race-proven details such as the upswept rear deck and sculptured side air intakes into a dramatic but uncluttered design.

A more streetable study was shown in 1968 called the P6. This was a mid-engined design and could have served as future inspiration for a production coupe. The curves were flatter, with details that were crisp and understated: all in the Pininfarina tradition of conservatism. At the Turin Show in 1969 Pininfarina exhibited a new and impressive design on a 512 S chassis which clearly demonstrated that coachbuilder's ability to interpret the squared-off wedge shape which had become the fashion on the race track. Pininfarina handled these elements in a masterful way and made the improvement look simple and obvious. The race-inspired "wedge look" was becoming a cliché among the other coachbuilders in Italy and with the 1969 512 S show car Pininfarina proved that they could bring to this motif a freshness that would be valid for future production car designs. In addition, in a seeming revolt against the wedge, Pininfarina brought forth the dramatic Modulo as an alternative. It succeeded in attracting crowds at the auto shows where it was exhibited but offered a collection of curves, shapes, negative and positive areas and holes that could not be assimilated into production cars, not even Ferraris. The Modulo answered questions that have never been asked.

On the production side, Pininfarina had not been idle. In 1968 three completely new models were introduced: the 365 GT 2+2, the 365 GTB/4 Daytona and the Dino 206 GT. Seating four people, the 365 GT 2+2 was clearly derived from the 500 Superfast and that model was dropped from the line.

The 365 GTB/4 berlinetta, referred to by everyone but the Ferrari factory as the Daytona, was destined to be the last front-engined twelve-cylinder two-passenger car made in Maranello. The design is conservative, yet the overall impression is brutish and purposeful.

8 365 GTB/4 by Pininfarina.

Even more attractive was the spyder convertible version which was built in much smaller quantities by Scaglietti—as was the coupe—until the Daytona series was discontinued early in 1974.

The Dino 206 GT coupe was Ferrari's first production mid-engined car and it gave Pininfarina's design studio an ideal opportunity to demonstrate their styling expertise with a mid-engined configuration. The somewhat modified production version available from the production line in 1968 was no disappointment. The design is uniquely Pininfarina and owes nothing to other influences. The well-rounded surfaces, the horizontal scoops on the doors leading rearward to air intakes for the brakes, the concave rear window with trailing buttresses on either side, all serve to contribute to a whole that has its own very special character. In the spring of 1969 the engine was enlarged to 2.4-liters and the model was renamed 246 GT. In 1970 a U.S. model was ready and importation into the States officially began. In 1973 a spyder version was created with a Targa-type removable roof and this replaced the previous coupe model. The Dino 246 GTS spyder continued in production until August, 1974.

In 1971 Ferrari introduced a luxury coupe which was only barely a 2+2 (rear seat room was very scant), called the 365 GTC 4. It was really a more comfortable, more tractable, long wheelbase version of the Daytona and its interior was luxuriously appointed and included individually adjusted reclining front seats. The exterior design was based on a minimum number of almost flat, intersecting planes. The nose terminated in a rather homely full-width bumper/grille unit. It was another model like the Lusso, indecisive and not too well accepted at the time of its manufacture. Production ceased at Pininfarina at the end of 1972 after less than two years.

Production of the 365 GT 2+2 ended in 1971 and its replacement was not announced until the Paris Salon of October, 1972. This new model was the 365 GT/4 2+2 and was a totally new car, a full four-passenger automobile, the largest ever built by Ferrari. The styling was crisp and rakish, using large, almost flat intersecting areas. The side glass areas are sharp and angular and the overall appearance is Daytona-inspired. Strangely, this model has never been imported into the United States.

At the Turin Automobile Show in 1971, an exciting new prototype was shown with a mid-engine chassis and two-passenger Pininfarina coachwork. The engine was a most impressive flat twelve-cylinder unit referred to as the Boxer. There was much speculation as to whether Ferrari might actually put this model into production and this was denied by the factory. Two years later the 365 GT/BB Berlinetta Boxer was officially announced and deliveries began. The design is pure Pininfarina with no visible, detectable outside influence, but again, Scaglietti does the actual fabrication. The gently rounded,

1968 P6 by Pininfarina.

flowing forms are understated, subtle, almost self-effacing. Indeed, the conservative quality of the styling is in sharp contrast with the capabilities of the automobile: it is the fastest production car in the world with a top speed of 187 mph.

By the middle of 1973 Ferrari had reestablished relations with Carrozzeria Bertone and for the first time since 1957 broke his exclusive relationship with Pininfarina for production car designs. Before the end of the year a new design was shown on a new mid-engined chassis called the Dino 308 GT 4, featuring a double-overhead-cam, transversely-mounted V-8 engine. The interior is of 2+2 configuration and the +2 portion is cramped albeit useable. The styling is from the wedge school but refined and with innovative air ducts on either side just back of the rear quarter windows. Its small size makes it even more appealing in reality than it appears in photographs. The U.S. version, approved and on sale in early 1975, uses squared-off black safety bumpers front and rear which blend with the basic design better than expected. Following Ferrari practice, the 308 GT4 is produced in the Scaglietti plant in Modena.

The addition of Carrozzeria Bertone as an official Ferrari body designer is a most welcome development. Certainly, Ferrari will benefit from the competition between the two great coachbuilding houses of Pininfarina and Bertone.

As this book goes to press in mid-1975, word of another 308 Dino has been announced, a two-place coupe by Pininfarina which appears to have been evolved from the P6 show car. It has a Boxer-like front and the rear resembles the 246 GTS, both elements blended smoothly to create the new form. It appears likely that this model will also be available with a removable roof. The new Dino is the latest example of Pininfarina's evolutionary approach to automobile design and to uniquely improving the breed. Of Pininfarina's work on Ferrari chassis, it might be said that their designs have stood the test of time. There is little more one could ask for.

Looking back on the nearly three decades since the first Ferrari was created, it is apparent that the variety of bodies gracing Ferrari chassis is unrivaled by that of any other marque, sporting or otherwise, of the postwar period. The best of these designs represent some of the most beautiful yet functional automobiles ever built. If some Ferrari bodies seem bland or plainly disturbing, it is probably also true that even these manage to convey the feeling that here is a special automobile conceived with a taste and style not always found even in other limited production cars. That there should be this dichotomy of the very beautiful and the occasionally bland is understandable for each and every Ferrari somehow reflects the individuality of the men who created it, and not everyone agrees on what is bland or truly lovely. Then too, of course, it is this very individuality, coupled with the cars' performance, which is what Ferraris have always been about. ✿

Ferrari in America

by Albert R. Bochroch

Luigi Chinetti in the pits at Le Mans, circa 1961.

A native Philadelphian now living in Bucks County, Pennsylvania, Albert Bochroch left a successful advertising career to devote full time to writing. His book, American Automobile Racing, *was published by The Viking Press in 1974 and he is currently working on a history of Americans at Le Mans. Mr. Bochroch's articles have appeared in several automobile publications including* AUTOMOBILE Quarterly, *and in 1968 he received an American Auto Writers and Broadcasters award. In this chapter, he draws on his long time involvement with motor sports to recount Ferrari's development in the United States—both on the track and in the showroom—and to tell the story of the man who for years has played an important role in this development, Luigi Chinetti.*

Winners of the 1951 Carrera Pan Americana, a 2000-mile, six-day blast along Mexico's rocky spine, were Piero Taruffi and Luigi Chinetti. That this veteran pair brought their 2652 cc Ferrari 212 Vignale Export coupe home first was not surprising. Each driver had captured more than his share of laurels in the classic sports car endurance races. Taruffi was soon to win both the Targa Florio and Mille Miglia, and Chinetti had been a three-time winner of Le Mans. And, as early as 1951, those who followed motor sports were beginning to expect a Ferrari to win.

What was remarkable was that Chinetti and Taruffi averaged 88 mph (10 mph faster than the winning Olds 88 in 1950) and that, in 1951, both drivers were over forty-five.

1949: Chinetti wins Le Mans.

A year later, Chinetti brought a Mexico 340 berlinetta to the Mexican road race—one of three entered by Ferrari—where he finished third to Karl Kling and Hermann Lang, both driving Mercedes 300 SL's. On the leg into Juarez, the last section of his last important race, Luigi Chinetti averaged 128 mph.

Although the Connecticut Turnpike carries the bulk of through traffic, the old Boston Post Road, U.S. #1, Putnam Avenue in Greenwich, still maintains a heavy flow. Chinetti Motors, on the right approaching Greenwich, is about the size of an average American car dealership. One minor difference may be the absence of a large used car lot. Only a few older Ferraris are lined up in front of the showroom, and several more are in the small parking area. Some of these, however, may be customers' cars.

Still it's unlikely that anyone short of Ferrari's own sales outlet in Modena could match Chinetti's dazzling assortment of forty-five to fifty Grand Prix, sports and vintage racing cars. The shop is an enthusiast's dream. Off to one side stands an immaculate silver Lusso, looking lonely, small and prim. Two of the new dart-like Boxers seem ready to fly. A customer's streaming GTB is surrounded by a small group that includes the worried owners, a young couple who fail to reflect the Gucci-Pucci Ferrari image. Standing aloof, a few feet away, is Luigi Chinetti. Managing to be quiet and nervous at the same time, he observes the scene. A salesman walks over. "It's alright, Mr. Chinetti," he says, "the owner forgot to replace the dip-stick. Some oil must have splashed on the headers."

If you are even remotely interested in Ferrari, if you only follow American racing casually, you know of Luigi Chinetti. You may have heard outlandish stories of customers bringing their cars in for service, only to be told to park them in a public lot—with the parking charge being added to their bill; or of Chinetti's refusal to service a car, or sell parts for it, unless he knew the car and the owner's pedigree. Shy, arrogant, soft-spoken, imperious, courtly, a Caesar—but which one?

Chinetti's Greenwich office, a small, dark and cluttered room, is really a museum. Photographs and Ferrari memorabilia line the walls. Side-by-side, three P4's thunder across the finish line at Daytona. "Luigi, I sure enjoy driving your Ferrari—Mario." On a file cabinet is a picture of two type 166's. A grinning Luigi and co-driver Jean Lucas sit behind the wheel of number 16, with a smiling, attractive young lady, Marion Chinetti, perched on the hood. In number 18 is André Pilette who finished second to Chinetti and Lucas in the Ore d. Pargi—1948 Montlhéry Twelve Hours. "To Luigi, best wishes— Steve McQueen." Inevitably, there are tragic photos of the fallen: Bandini, Rindt, Scarfiotti and the Rodriguez brothers, for whom Chinetti still grieves.

Dick Fritz, whose job as team manager of NART, the North American Racing Team, appears to include being Chinetti's "Man Friday" in dealing with supplies and customer relations, comes in to discuss a problem. As he leaves the office, he turns and asks, "What other shop would give a new car guarantee on a used car, and then give the customer his money back after two months?"

A large painting of a Bugatti by Géo Ham carries a long inscription. Nearby is framed a "Diplome de Citoyen de la Ville du Mans, 1972 Juin, Luigi Chinetti, 1932-49." A shop man enters and asks if he can sell a part. Luigi Chinetti wants to know who the customer is. Only when he is convinced that it is someone they know does he say, "Okay, but no discount." There are Ferrari owners who have found Chinetti's parts-selling and pricing policies frustrating and disorganized and ultimately humiliating.

The room's most conspicuous object is a prancing horse in black iron, mounted on a red Carrara marble base. The inscription, in raised letters, reads "North American Racing Team 52 53 54 56 57 58 60 61 62 63." We learn that the statue was a gift from Enzo Ferrari, given at a time the Commendatore and Luigi Chinetti were close.

There are more than casual similarities between these two strong-willed sons of the Savoy. As you talk, it becomes obvious that Chinetti's feelings toward Enzo Ferrari are ambivalent. In his autobiography, published in the U.S. in 1964, Ferrari almost ignores Luigi Chinetti. Only in connection with his having arranged for Adlai Stevenson to meet Ferrari is Chinetti mentioned. The Commendatore says: "His arrival, accompanied by his wife, son and daughter-in-law, was announced to me by Luigi Chinetti. Chinetti is still working with me as he has done since the old Alfa Romeo days. He drove a Ferrari to victory in the 1949 Le Mans 24-Hour Race and the Spa Grand Prix." These are Enzo Ferrari's only references to Luigi Chinetti in the entire book.

Are Americans overestimating Chinetti's importance to Ferrari? Or is Luigi a victim of old-fashioned European chauvinism—of the notion once widely held that American racing lacked quality? Is it proper for Ferrari to so dismiss his lifelong friend; the driver who first brought Ferrari international laurels by winning the 1949 Le Mans; the business colleague of a quarter-century who is regarded as responsible for the Prancing Horse being on other than race cars; the agent who once sold such a large share of Ferrari's entire production; the patron of the North American Racing Team, whose NART Ferraris have graced so many victory lanes?

Both men have devoted their lives to a race car. But only the builder has it bear his name. Does Enzo Ferrari envy the Chinettis? Ferrari lost his only child, the beloved Dino. The Chinettis have Coco, their handsome son, and he is very much a part of Chinetti Motors.

**Bridgehampton, 1949: George Rand
with Alfred Momo and friend.**

Although not in the same league as his father as a race driver,
Coco—Luigi, Jr.—has had 5th and 13th place finishes at Le Mans. In
September of 1974, Graham Hill, Milt Minter and Coco drove NART
Ferrari coupes to a slew of international records at Bonneville. One of
Coco's solo runs in the 512 MB, 174.759 mph for ten kilometers, was
an international class C record and the session's fastest individual
effort.

Few Americans see the international racing scene as astutely as
René Dreyfus. On the subject of Chinetti's place in the Ferrari sun,
René says, ''You have to admit that Luigi Chinetti is responsible for
Ferrari's success in the United States. And, because of what Luigi was
able to do for him here, Ferrari was able to go on and do many things
he may never have been able to.''

Luigi Chinetti talks about Enzo Ferrari only with reluctance.
''Some day when I retire,'' he smiled and pointed to the drawers of
his desk, ''more than I'm retired now, I will write my own book about
Ferrari.'' Does he see Ferrari? ''When I go to Europe, I see him, but
it's Mrs. Ferrari I like to visit . . . she has had a hard time, but she is a
very fine woman . . . please say something nice about her.''

Born in Milano in 1905, Luigi attended the local Polytechnic for
two years before working for his father, a mechanical engineer. ''I
wanted to go right to work for Alfa Romeo, but my father said I must
first learn something about his business.''

Chinetti began racing in 1928, and by 1932 was a member of the
Alfa Romeo competition department. Contrary to legend, Chinetti says
Raymond Sommer did not drive the winning Alfa all but three hours
in 1932: Chinetti claims that he himself drove at least ten. Chinetti
also says that the story of Lord Selsdon having overindulged in 1949,
the year Chinetti won his third Twenty-Four Hours and a Ferrari its
first, is false. Yes, says Luigi, the Englishman drove for only four or
five laps, but it was not because he was drunk.

Mrs. Chinetti, a handsome, well-dressed woman, who was in her
husband's office at this time, said, ''He was not used to the wine, and
it did not agree with him . . . I was there and so was Coco—he was
two years old . . . It was the year I got an eye infection, and it still
bothers me . . . Lord Selsdon was a big, strapping fellow . . . he did not
feel well, but he was not drunk.''

''Look how well and how long he drove the next year,'' said
Luigi. Selsdon and Jean Lucas did drive the same two-liter Ferrari 166
MM in the 1950 Le Mans, going out Sunday morning while lying
sixth after being in an accident.

By the mid-Thirties, Italy was an uncomfortable place for those
who disagreed with Mussolini. Strongly anti-Fascist, Luigi Chinetti got
out. He roamed about Europe, spent some time in France and settled
in Paris. In the spring of 1940, Luigi Chinetti was hired by Mrs.

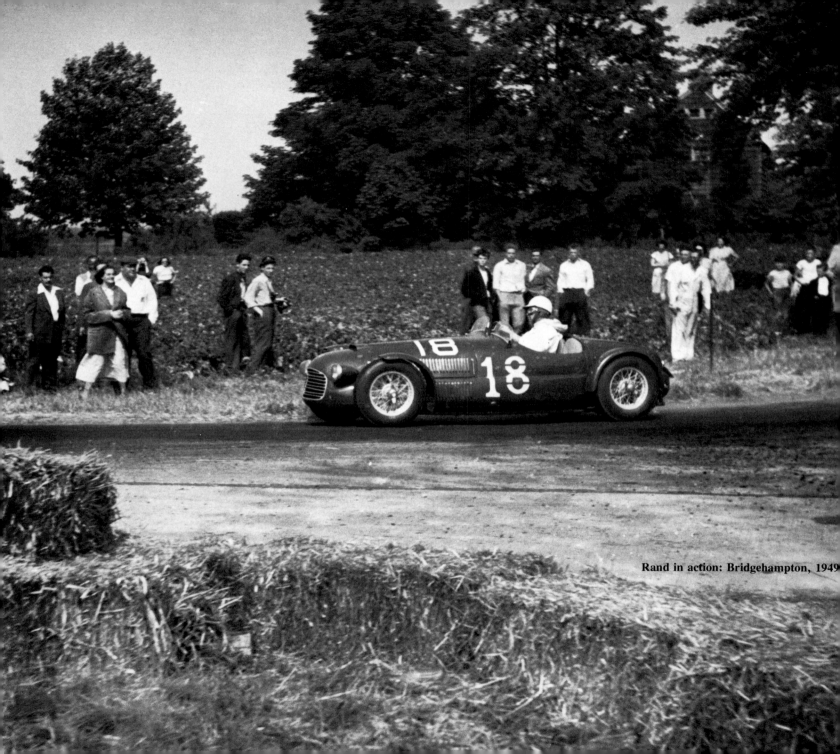

Rand in action: Bridgehampton, 1949

Laury Schell as chief mechanic for her Indianapolis cars, the Lucy O'Reilly Schell Maserati Specials. The drivers were René LeBegue and René Dreyfus. Madame LeBegue and the Schell's nineteen year-old son, Harry, completed the party.

By this time, Germany had invaded Poland and France was at war. It was the time of the Maginot Line, of the "phony war." Officially, America was still neutral and the 500 was to be held in 1940 and 1941. Although both René Dreyfus, by then champion of France, and René LeBegue were serving in the French army, they were able to get leaves to race at Indianapolis.

René Dreyfus, who has remained a close friend of Luigi Chinetti and of his family for many years, recalls the group's American adventure. "We came over on an Italian ship...Even though Italy had not joined Germany, the boat was full of rumors that she would soon declare war...Here we were, two members of the French army and an Italian who was known to be against Fascism...It was a very nervous time, and we hoped that once we got in American waters, they could not try to keep us on board . . . The only one of us who could relax and enjoy himself was young Harry Schell...He was an American citizen and America was not at war...We landed on May 23rd and were taken immediately to the World's Fair in Flushing where we had our pictures taken at the Ford Pavillion...Then it was off to La Guardia and we got to Indianapolis that same evening."

Short on practice time and handicapped by the need to change a broken engine, the French pair only qualified Maserati number 49 in the thirty-first spot. The second Lucy O'Reilly Schell Special was bumped after making the field. Sharing the ride, the two Renés moved up steadily, bringing number 49 to a solid tenth place finish.

The "phony war" ended as German troops swept into Belgium and France. Over the Memorial Day weekend, from May 26th until June 3rd, 350,000 English and French troops had been evacuated from Dunkirk. Italy joined the Axis on June 10th, and on June 14th Paris fell.

It was a difficult time for the entire group, but especially so for Chinetti. The team could not return to France, and none of them were fluent in English. Eventually the two Frenchmen were able to join the American army. Although Chinetti was known as an anti-Fascist, he was considered an enemy alien by the U.S. government. Luigi was able to obtain work as a mechanic in New York, but his movements were restricted, and not until 1946 was he able to obtain his American citizenship.

During his early days in America, Chinetti met Marion, his future wife. She was of Belgian descent and, helpfully, spoke fluent French. Before going into business for himself, Luigi worked for Alfred Momo as a mechanic at J. Inskip's Rolls Royce agency.

Chinetti and Ferrari had first met when Luigi joined Alfa Romeo. But it was 1946, when Luigi returned to Europe to clean out his Paris garage, before they were to meet again. Yes, Luigi says, the story is true, it was his suggestion that started Ferrari building cars for customers. The 166 Spyder Corsa sold to Briggs Cunningham in 1948 was Luigi's first American sale. And Chinetti with the Collier brothers and Alec Ulmann started Briggs Cunningham thinking about Le Mans.

The October 1948 Watkins Glen uncorked a healthy U.S. interest in sports cars and in road racing. Inspired by the prewar ARCA, (Automobile Racing Club of America), the road racing renaissance, forced to lie dormant during the war years, now exploded. It was a good time for Luigi Chinetti to start selling Ferraris in America and the U.S. Ferrari market would eventually account for some twenty-five to thirty percent of the touring car production. Still, it has not been an easy road and there have been times when Chinetti needed a customer's check in order to get that customer's car off the docks. There have been several silent partners in the agency as well, something which has kept the business going at times when it might have otherwise failed. Chinetti's best year was 1965-66 when some 240 cars were sold. He initially opened for business in a rented 49th Street garage in New York City early in 1948 but, by the end of the year, moved to a loft on 19th Street. It was not until 1951 that he occupied his long-time headquarters on 54th Street, near 11th Avenue. Although Greenwich was opened in 1965, it did not become Chinetti Motors' main office until 1971 when the 54th Street premises were closed.

An Alfa Romeo had captured that first Watkins Glen of 1948, and another Alfa won on the four-mile Bridgehampton street course the following June. But at Bridgehampton George Rand drove the Cunningham Ferrari to a big lead—a hint of things to come—before dropping out with a broken oil line. The second Watkins Glen Grand Prix, increased to ninety-nine miles in 1949, but again held over the same 6.6-mile combination of village streets and country roads, some of them unpaved, was taken by Miles Collier's Ford-Riley with Briggs Cunningham second in his 166 Ferrari Spyder. Cunningham's Bu-Merc—a Mercedes SSK body on a 1936 Buick Century chassis powered by a slightly tuned straight-eight Buick engine—was third. Tom Cole drove an HRG to fourth, a lap ahead of John Fitch in an M.G. Earlier, Cunningham had driven the Ferrari in the four-lap Seneca Cup, finishing second to George Weaver's single-seater Maserati, with Sam Collier's supercharged M.G. a close third.

The first Elkhart Lake race, run in July 1950, over 3.35 miles of open roads on the northwest side of the lake, was a victory for Chicago SCCA stalwart Jim Kimberly in his 4101 cc Ferrari 340 America. Not until 1955, after the State of Wisconsin banned racing

Charles Moran finished 9th at Watkins Glen in 1951.

on public roads, would the park-like Road America course be opened.

The 1950 Watkins (the locals snort that only tourists call it "the Glen") was held on September 23rd and three Ferraris were among the more than one hundred sports cars entered in the Queen Catherine Cup, Seneca Cup and Grand Prix. Sam Collier was to drive Briggs Cunningham's 166. Gentleman Jim Kimberly and Big Bill Spear would be in new 166 Mille Miglias.

During that Grand Prix, Samuel Carnes Collier, age thirty-eight, was killed early in the race when he went off the road on the high speed bend beyond the old railroad underpass. The GP was won by Erwin Goldschmidt's Cadillac-Allard. Cunningham was second in his Cadillac-Healy.

A quarter century has passed, but should you visit Watkins Glen, chances are that you will find fresh flowers on the small stone marking the site of Sam's accident.

The early 1950's proved to be a time of newly developed hot cars and fierce driver rivalries in U.S. road racing. Brutal Cadillac and Chrysler J-2 Allards won at Bridgehampton, Santa Ana, Pebble Beach and Reno. In March 1951, George Rand led a group of Americans to Buenos Aires for the General Peron Grand Prix where Fitch drove Tom Cole's old Allard to defeat the Spear, Kimberly and Cunningham Ferraris. Rand, who drove Cunningham's Ferrari 166 in the Argentine, recalls that Briggs' first reaction to Sam Collier's death while driving his Ferrari had been to sell it. He did not want to see it again. One reason Rand took the 166 to South America was to find a buyer. But no one was interested, and Ferrari No. 016-1 is today on display in Cunningham's Costa Mesa museum.

Until the Cunningham-supported D Jags came along late in 1955, Ferrari versus Ferrari was the big attraction in SCCA nationals, the premier U.S. road races of their day. The Kimberly-Spear matches developed regional followings; red-suited Jim representing the Midwest, and Connecticut's Big Bill the eastern establishment. Spear's Ferrari won the 1953 Sports Car Club of America National Championship, a questionable term perhaps as the distribution of events favored east-of-the Mississippi drivers. In 1954, when the use of Strategic Air Command bases temporarily solved U.S. road racing's perennial "where to race" problem, Kimberly's new 4.5 liter Ferrari 375 MM won at MacDill, Hunter and Watkins Glen. Spear, with wins at Andrews in May and at the March Air Base in November, was close, but Kimberly took the SCCA 1954 C Sports Racing title, as Bill Lloyd's Ferrari captured the 1954 D Modified crown. Some idea of Ferrari's popularity at this time is found in the March Air Base race results where Ferraris finished 1st, 2nd, 3rd, 4th, 6th, 7th and 10th.

Wealthy American enthusiasts Temple Buell, Tony Parravano, John Edgar and Allen Guiberson provided Ferrari rides for Hill, Gregory,

Phil Hill at Pebble Beach, 1951.

Bill Spear at the 1951 Watkins Glen GP.

Spear's 4.1 liter Vignale-bodied Mexico: Bridgehampton, 1953.

Phil Hill at Hagerstown, 1955.

Torrey Pines, 1955: Ernie McAfee's 121 LM.

verly Airport, Mass., 1956. Jack McAfee and a 3.5 Monza.

rroll Shelby won the 1956 Beverly Airport SCCA National in this 4.4.

Shelby and the McAfees, Ernie and Jack. Richie Ginther worked as salesman and manager for California Ferrari agent-race driver Johnny von Neumann. Before they became members of the tradition-encrusted European factory teams, America's young masters, Phil Hill, Dan Gurney, Richie Ginther, Carroll Shelby and Masten Gregory, learned much of their trade driving Ferraris in the United States.

Early Ferrari sports cars were amazingly durable. Phil Hill's first Ferrari, a 2562 cc V-12 roadster, had already won the 1951 Tour de France and had run the 1951 Le Mans by the time he bought it. Then Phil finished second at Golden Gate and Pebble Beach and won Torrey Pines. When Hill loaned the car to *Road & Track* for a road test in 1952, the engine had never been opened. His second Ferrari was a 2900 cc Vignale roadster. By 1955, when he became the SCCA D Sports Racing national champion, he was driving his fourth Ferrari. With it Hill nosed out Sherwood Johnson's new D type Jaguar to win

the first Road America, but Johnson went on to take Watkins Glen and win the 1955 SCCA C Modified championship. For the next four years, C Modified was dominated by Walt Hansgen as Ferrari drivers James Johnson, Alan Connell, E.P. Lunken, Gaston Andrey and Tom O'Brien monopolized smaller displacement modified honors.

Ferrari's domination of U.S. sports racing classes began to fade in 1956 when Walter Hansgen started driving superbly prepared Cunningham Jaguars at a time other "world class" American drivers were busy overseas.

Whatever the problems in sports racing classes, Ferrari GT cars fared well in the U.S. from 1958 to 1961, and annual SCCA national production car laurels were won by Ferrari coupes in the hands of George Reed, Bob Grossman, Robert Hathaway and Charley Hayes. Phil Hill, Masten Gregory, the Marquis de Portago, Stirling Moss and the Rodriguez brothers were among the Ferrari drivers

Luigi Musso in the 290 MM he and
de Portago drove to 7th at Sebring in 1957.

winning feature races in Nassau during the annual Bahamas Speed
Week.

It was 1964, incidentally, which saw the rather extraordinary
appearance of Ferrari Formula One cars entered in the U.S. Grand
Prix not by Enzo Ferrari but instead by Luigi Chinetti. The cars were
painted blue and white—NART's colors—rather than the traditional
red and carried NART emblems. Behind the changes, of course,
loomed Ferrari, once again embroiled in one of his many disputes with
race organizers or rules makers.

This time, Ferrari was displeased with the A.C. d'Italia. The Club
had assured Ferrari that there would be no problem in homologating
the 250 LM as a GT car. Then the Commission Sportive Internationale
refused the application on the basis that the Club had not presented
sufficient evidence to show that the necessary one hundred cars had
been built.

Ferrari saw no valid reason for the decision, believed the Club had
handled the situation poorly and announced that he was going to resign
his membership. In addition, he said that after Monza his cars would
no longer run as an Italian works team. Hence their novel color
schemes at Watkins Glen. Surtees finished the 1964 U.S. GP a strong
second in a V-8 car: Bandini's flat-twelve broke. Ferrari's disputes
with the C.S.I. continued, of course, and in 1965—after a problem
emerged with homologating the 275 GTB—Ferrari issued a statement
saying that he was now withdrawing from the Championship of
Manufacturers. But that's another story.

By the early Sixties, the Ferrari sales and service organization in
the United States had been substantially expanded when Bill Harrah
became a dealer. Harrah—impressed after a fast ride in a Ferrari
driven by Richie Ginther—added the marque to his Modern Classic
Motors, established in 1959 to sell Rolls-Royce. He paid Luigi

183

George Arents' 500 TRC at Sebring, 1957.

Chinetti a royalty on each car sold until 1969 when he acquired the western area distributorship from Johnny von Neumann.

Harrah's Ferrari organization has boosted sales in the western U.S. to over two hundred cars per year. Equally important is the reputation that the Reno-based organization has garnered among Ferrari enthusiasts. "We've tried to bend over backwards to build our image as reliable and honest," Vernon C. Keil, the vice president and general manager of Modern Classic Motors told me. These efforts have been largely successful, refreshingly so considering the number of men with reputations as scoundrels and dirty dealers who have for years preyed upon Ferrari owners and would-be owners in this country.

In addition to Harrah's organization, Al Garthwaite's Algar Enterprises also serves as a Ferrari distributor. Based in Paoli, Pennsylvania, near Philadelphia, this branch of Garthwaite's business is known as Chinetti-Garthwaite. It has been in operation since 1965.

Whatever the changing face of the business organization in the U.S., racing efforts have remained largely under Chinetti's auspices. During the mid-Sixties when Ford began giving Ferrari competition in the Manufacturer's Championship and won at Daytona, Sebring and Le Mans, Chinetti's NART 250 LM—driven by Masten Gregory and Jochin Rindt—won Le Mans in 1965. At Daytona, Pedro Rodriguez twice brought NART victories in the 24 Hours and in 1967 factory P4's raced to an inspiring one-two-three, three-abreast finish. Ferrari works entries also captured Sebring eight times between 1956 and 1970.

When the Sports Car Club of America introduced its first professional road racing series in 1963—the Can-Am—the need for big bore engines found Ferrari unprepared or unwilling to compete. Not until 1969 when the 6200 cc V-12 Can-Am car was built did Ferrari seriously respond to Group 7 racing's almost universal dependence on Chevrolet V-8 power. It was a response which met with indifferent results. In 1967, Chris Amon drove a modified P4 in the season's three final Can-Ams, fifth at Monterey being his best showing. Pedro Rodriguez was fifth at the 1969 Bridgehampton Can-Am in the NART 312, and Amon returned in 1969 to give the Bill Harrah-supported 612 some brilliant drives. But thirds at Mid-Ohio and Watkins Glen, and a second at Edmonton were the best finishes for Ferrari Group 7 cars.

The Can-Am performance perhaps pointed out a facet of the Ferrari-Chinetti relationship which should be mentioned. That is, Ferrari has probably always viewed U.S. racing with a certain disinterest. "He doesn't seem to realize what racing means to us here," Dick Fritz said, "because he doesn't give us all that much support. We have to really keep after him about this. I've had to ask

185

250 Testa Rossa

Mrs. Ferrari personally if we couldn't have certain racing cars over here. We've got the cars but we've never really impressed the importance of seeing Ferraris race in the U.S.—its effect on sales and image—to any tangible degree."

This situation has been reflected in Ferrari's lack of interest— since 1952 at least—in Indianapolis too. Ferrari agents and Ferrari drivers have all shown more enthusiasm about running Indy than Ferrari himself and their efforts are recounted by Karl Ludvigsen elsewhere in this book. In the spring of 1975, the United States Auto Club reported that "Dan Murphy expects to have as many as three Ferrari-engined Cicadas at Indianapolis with a view to running two in the 500 mile race . . . The engines are normally aspirated five-liter 512 sports car powerplants and will be reduced to 274 cubic inches to meet USAC requirements." The report proved overly optimistic, however, for no Ferrari powered cars appeared at Indy in 1975.

It was traditionally in the classic endurance races, particularly Le Mans, where Chinetti controlled hard-to-come-by invitations, and NART's "buy-a-ride" policy became part of many Yank challenges in the Twenty-Four Hours. Surprisingly, Chinetti was not reserved in talking about the business of "buying-a-ride." How did he decide who bought a car? Why was one driver given a ride, while another driver would have to buy his North American Racing Team entry? The answers were shorter than the questions: "If he is a rich man, he pays. I don't care if he is the best driver in the world; if he is rich, he pays."

One of Chinetti's more successful "buy-a-ride" customers was car dealer Bob Grossman. Grossman's Nyack showroom doesn't exactly match Chinetti's as an exotic car emporium, but it comes close. In addition to a score of Maseratis and Lamborghinis, which are expected, as Grossman is the eastern distributor for both Italian delicacies, there are several lines of current English cars, and an assortment of rarities which, on my visit, included a Mercedes Gullwing and sundry used Ferraris.

We talked with Grossman in his eagle's nest of a home above the Hudson. Recalling that he had attended the Philadelphia Museum School of Art (he also studied voice and briefly sang professionally) softened the surprise of seeing the brilliant art on his walls. Even more unexpected were the names, which included Matisse and Chagall, and the quality. "Well," Grossman explained, "some of them were swapped for cars.

"I came up here in 1946 after the army and opened a garage . . . Max Hoffman began selling me Jaguars, but only if I agreed to take one VW with each three Jags . . . For a while I raced Jags and Alfas, but I wanted something faster and started driving Ferraris." (Grossman won the SCCA G Production title in 1958 in an Alfa Romeo, C Production honors in a Ferrari in 1959, and tied for the A Production

February, 1967. Daytona winner Chris Amon in the pits.

championship in 1961, also in a Ferrari.)

Grossman drove the Twenty-Four Hours nine times between 1959 and 1973. He finished in the top ten seven times, five of the seven being in a Chinetti NART Ferrari.

"Only twice in all those years," Bob went on, "did the deal work out the way it was supposed to. In 1959, Luigi said he'd get me a California convertible similar to one I had liked, only the factory would set this one up for Le Mans with aluminum panels and a stronger engine. I drove with Tavano, the stone mason from Le Mans, and we finished fifth; and just recently, in 1971, Coco Chinetti and I came in fifth again in a GTB . . . Except for those two races, it was complete confusion; you never knew where you stood until a half hour before the race started . . . It wasn't enough that you bought your own ride, although it wasn't a bad deal. Until just before the race you did not know who you were going to drive with, or if you'd drive at all. It was just chaos.

" But I still have a lot of respect for Luigi. You can't help feeling for a guy who has such a love of cars, and he's one of the shrewdest men I've ever met in the automobile business. He's a fox, yet he can be absolutely charming, especially when he's in Europe. I really used to enjoy eating with him in Le Mans. We would talk about wine and food . . . Overseas, he seems like a different person. You know, Chinetti has a wonderful knack in picking good new drivers—he's found a lot of them. But he sells cars to people who have no business driving at Le Mans."

What of Ferrari's future in America? The future would be dim if Ferrari sales depended, as they historically have, on racing publicity. But for Formula One, Ferrari has little to talk about. The International Manufacturer's Championship is in the doldrums. Except for an occasional stray, the prancing horse emblem is seldom seen in U.S. club racing—other than the vintage variety. Then too, the "times"—inflation coupled with recession, fuel shortages and a national speed limit that extends even to the once promised land of Nevada—create a poor climate for selling $30,000 cars. In fact, the Dino 308 has not sold at all well in this country.

The future then, is as hard to predict in the U.S. as it is in European markets. What does seem to be happening here is an ever increasing interest in restoring and collecting older Ferraris, both racing and touring models. At this point, the U.S. Ferrari world is composed of a peculiar blend of people which includes those who could be termed nouveau riche and who buy Ferraris for the car's image and chic though they are not necessarily enthusiasts; on the other hand are people who love the cars for what they are—good investments be damned—and view their Ferraris as a hobby. In times of uncertain values, men who can afford it turn to precious playthings. ⊕

CHAPTER 8

Ferrari, Ferraris & Me

by Stirling Moss

Before his brilliant racing career was cut short by a crash at Goodwood on Easter Monday, 1962, Stirling Moss had become recognized as one of history's greatest racing drivers. Since his retirement from racing, he has turned to various business interests and co-authored two books, Design and Behaviour of the Racing Car *with Laurence Pomeroy and* All But My Life *with Ken Purdy. He had earlier written* In the Track of Speed *and* A Turn at the Wheel *and has long been a contributor to various magazines including* AUTOMOBILE *Quarterly. Here, Mr. Moss relives his relationship—or lack of it—with Enzo Ferrari and, in doing so, poses one of racing's most fascinating might-have-been's.*

I think I'm safe in saying that, of all modern constructors of racing cars, it is Enzo Ferrari who most identifies himself with his creations. Whether he does the bulk of actual design work or not matters little. His is the driving force behind the machines, the sensibility which makes the cars what they are. To me, there has always been a certain special beauty or extra style to his cars which makes them unique. A large part of this is doubtless the part they have played in motor racing history and their association with so many great names of the past like Alberto Ascari, Luigi Villoresi, Froilan Gonzalez and, to my mind the greatest of all, Juan Manuel Fangio. It's also the memory of Gonzalez storming round Silverstone to break the 100-mph practice lap barrier for the first time and then forcing

Running hard at Silverstone in a SWB Berlinetta.

Fangio to settle for second place in the 1951 British GP. Or Taruffi thundering home to win the very last Mille Miglia. Or Mike Hawthorn winning the World Championship in a contest that—as I know only too well—was in the balance until the last seconds of the last race of the season.

Living with a genius is supposed to be difficult, and living with Ferrari, who is absolutely single-minded in pursuing an ideal, can't be easy. I came up against the harsher side of Ferrari's thinking early on in my career. I'd been racing for just over three years, starting with a Cooper 500 in British events in 1948 and following this up with another Cooper season in 1949, this time driving in Continental races as well. By the end of the 1950 season, I'd driven Coopers and HWMs all over Europe and, within the limits of the cars I was driving, I'd done reasonably well. So I was pleased and flattered when I heard that Enzo Ferrari was going to offer me a place on the works team for the Formula Two race at Bari at the beginning of September 1951.

Yet when I turned up for practice there, I had the shock of my life. I asked which was to be my car, and I was coldly informed that there was no car for me. The Commendatore had changed his mind, it seemed, and that was that. Thus the offer was withdrawn and the car which I would have been driving was given to veteran Piero Taruffi instead—no apology, no explanation. Now I'd got nothing at all against Taruffi, but the abruptness of the decision, and the matter-of-fact manner in which I was told about it, had really got my back up. I went off swearing that Ferrari had seen the last of me for good.

In fact, I did drive a Ferrari in that Bari race. I borrowed a two-liter car from a chap called David Murray, but this proved to be a bad move. Not only was it the first real foreign racing car I had driven, but it had the old centrally-located accelerator pedal which I had never met until then. It proved quite a handful. It was all right so long as I could concentrate on what was happening. I could remember to reverse the normal pedal positions in my mind and cope with things without losing any time. But as soon as something happened, and I had to act automatically, then I forgot and it was all too easy to hit the throttle instead of the brake pedal. I had a slight shunt in the end, something I couldn't help, but enough to put me out of the running.

And sure enough, that Bari race was the last I did see of Enzo Ferrari, at least as far as driving his cars was concerned, for many years. In 1954, I raced my own Maserati for a season and after that I joined the Mercedes team and drove for them until they withdrew from racing following the disastrous crash at Le Mans in which one of their cars was involved. Following this, I went on driving Maseratis and Aston Martins, Vanwalls and Coopers and Lotuses but never—for a

long time—any Ferraris. On the other hand, I was almost always competing against them, and the more I had to try and beat them, the more I came to respect them.

As one who competed against Ferraris for many years, I came to realize that, although an incredible variety of designs has been turned out by the Maranello works, they invariably had one particular virtue in common, their reliability. Ferraris are often on the heavy side by the standards of other racing machines, but the power of their engines often means that they can afford the extra weight. Ever since the first appearance of the classic V-12, Ferraris have had plenty of power. And it's entirely characteristic of Ferrari, the man, that he uses this advantage to build extra strength into his cars. For he is on record as saying that to be beaten is one thing, to crash is another, but for a car to break down under the sheer strain of racing is the ultimate disgrace. Ferraris are tough because Ferrari himself would have them no other way.

Nothing matters to Enzo Ferrari as much, I believe, as his cars winning races. Anyone who works for him or with him must learn to accept this first of all. This drive to win is perhaps one reason why many people feel that human beings matter little to Ferrari. It has been said that Ferrari seems to regard his drivers as something of a means to an end, as important parts of the cars themselves rather than individuals in their own rights. Perhaps there have been instances when this has really seemed to be the case, but I don't think that what many of Ferrari's critics have said of him—that he has no respect for drivers—is true. I feel he does regard his drivers with respect just so long as he believes they are giving their best. I don't think it is even entirely a question of a driver's ability, although one does need a fair measure of ability to find oneself driving for Ferrari in the first place. Enzo Ferrari doesn't suffer mediocrity gladly, but with him attitude is very important. He is never satisfied with less than your ultimate performance, whatever the conditions.

Ferrari's attitude about his drivers is decidedly individual in various ways. For instance, there have been plenty of times when he's had more drivers on his books than cars for them to drive. I don't know for certain whether or not this is deliberate policy, but I think Ferrari's organization could field two cars for every team driver if Ferrari wanted to. In the 1957 Monaco Grand Prix, for instance, three works cars were wrecked in crashes, and yet by the next event on the calendar, the French Grand Prix, a set of brand-new replacements was available.

Another most important part of Ferrari's dealings with his drivers is, of course, his practice of not naming a team leader. When each driver feels he has an equal chance of victory and championship points, competition is that much keener. Unfortunately, drivers don't usually respond all that well to this kind of pressure. Ferrari's magic has attracted—and continues to attract—most of the world's top drivers, but they all seem to have grown dissatisfied sooner or later and left.

Another part of the problem of driving for Ferrari may be that when one is a member of such a high-class team, one is especially open to criticism. If you have been provided with what should be the best racing car in the world, and you're not winning with it, then the obvious conclusion will be all too easy to make—there is something wrong with your driving, or you're not trying hard enough. If you're struggling with a mediocre car, you're not so susceptible to this sort of thing and critics can't assess your form or ability so easily.

I myself was especially lucky in my early days, when I was driving for John Heath's HWM team, in that I didn't really have the kind of pressure one would feel as a driver for Ferrari. I don't mean that the HWM was a mediocre design but that the group was a small one without the backing enjoyed by the big works teams. Expecting the

HWM to be really competitive with the best GP cars of its time was clearly hopeless. So nobody anticipated that we'd do very well and the result was that, when things did go right for us, everyone said, "That's really great." If we were off form, though, nobody tried to say we hadn't been doing our best because then we were only doing what was expected of us. As a driver, then, I shared the best of both worlds at HWM. And people were always willing to give me and the team the benefit of any doubts.

Not so at Ferrari. Anyone driving for Enzo Ferrari needs to convince the boss he's doing his best all the time, and that's not easy. Nor is it easy to convince the racing public, either. As I've said, being at the top of the tree, one is expected by others to perform. And human nature being what it is, critics sometimes watch eagerly for the first signs that you are falling from your privileged position. Of course, nobody, not even Fangio, could win all the time. Even in the world's best car, racing is far too uncertain and unpredictable a sport for that. After my years in racing, I remember instances when even

Pit stop, Goodwood TT, 1960.

one or two failures against unusually determined opposition was enough to make people say of someone, "he's lost his touch" or that he, or his car, were no longer what they had been. In all, then, driving for Ferrari has never, ever, been easy.

Basically, however, I must say that I'm in agreement with most of Ferrari's thinking about how to handle his drivers, including not naming a team leader. Take, for example, the 1974 World Championship where you had drivers often basing their tactics on the way in which the series was going rather than simply going out there and tackling each race as a contest in itself. At that year's U.S. Grand Prix, for instance, I didn't feel that all the drivers were trying all that hard to win the race. Emerson Fittipaldi, I think, did the intelligent thing in trying to clinch the Championship, which did not necessarily mean having to win the race. I'm not trying to knock what he did; under the circumstances, it was perfectly understandable. But I do feel it's a shame when racing stops meaning simply trying as hard as you possibly can in every race on every lap. I've always believed that this is what racing is all about, and Enzo Ferrari knows this, too. He

understands and sympathizes with this kind of thinking.

And this is why the relationship between Enzo Ferrari and myself began to change, gradually but definitely, over the years. Although I never drove a Ferrari in a race for more than six years after Bari, it became increasingly obvious that Ferrari and I saw eye to eye as far as our attitudes towards motor racing were concerned.

My racing philosophy was very simple, really. I would rather lose a race by driving fast enough to win than win by driving slowly enough to lose. In other words, the important thing was to try as hard as possible—to me the word "race" means "Get in there and have a dice." Whatever the situation, whatever the car. There were occasions, of course, when taking care of a sick car might make all the difference between finishing and not finishing. But—in its way—even this involves doing your best within the limits of that particular situation.

In this respect, Ferrari and I were very alike. We were both trying to do all we could to realize our own personal ambitions. I was trying to drive as fast as I possibly could on the track while he was building

Pit Stop, Goodwood TT, 1960.

cars in the best way he knew, to make them go as fast as they possibly could. That we were both of very much the same mind was brought home to me in 1958 after Mike Hawthorn won the World Championship in his Ferrari. It had been a close fight all season between the two of us and, at the start of the season's last race, the Moroccan GP, there was still a chance of my taking the title from him if I could win the race, put up the fastest lap, and also ensure that he finished in no better than third place. In the end, it was a real cliffhanger finish. I got ahead of Mike in my Vanwall; I took the fastest lap; and at the end of the race I was clearly going to win, with him in third place. Then, in the closing moments of the contest, Mike's team-mate Phil Hill, who was lying second to me, spotted what was happening. He dropped back into third and let Mike through to take the title by a single point!

I was absolutely heartbroken at the time. Adding to my dejection was the fact that earlier that year I had given evidence on Mike's behalf after the Portuguese Grand Prix and it changed his disqualification into a points-winning second place.

A simple bit of tactics by Phil Hill had decided the outcome of a contest for which Mike and I had been driving our hearts out for a whole season. But I certainly don't hold it against Mike or Phil: it was the correct thing to do.

All the same, Ferrari himself understood how close things had been. He gave Mike a super gold medal with the Ferrari prancing horse beautifully enameled on one side and, on the other, the Union Jack and the Italian tricolor united. He had another medal made and gave it to me, a gift which came as a great surprise and which I am proud to possess to this day. I am sure he gave it to me because he realized that it takes two to make a contest, and, although I wasn't driving for his team, it was often me whom Mike was trying to beat.

But an even greater surprise occurred almost four years later. In 1962, I went to see Ferrari at his factory and he showed me all over the plant. He let me see all the cars and even gave me a look at his plans for the following season's Grand Prix races. He asked me to drive for the works team and said that I could have any car I wanted, embodying all my own ideas. He told me that he would build it for me in just six months, a promise which meant more to me when it came from a man like Ferrari than from anyone else.

It was an incredible offer, and it gave me a very difficult decision to make. I had always tried to drive British cars whenever I could and I didn't really know exactly how I would fit in as part of the Ferrari team, either. I had always found them—from designers like Chiti down to the mechanics—to be friendly and helpful opponents. But actually joining the team was different. I can remember asking Fangio once, ''Should I drive for Ferrari?'' And he answered, ''Certainly,

Ferrari cars are fantastic. Drive for him but don't sign a contract. Try and stay a free agent if you can.''

One thing I believed was certain: As Ferrari and I believed very much the same things about racing, we would get along. We both wanted to thrash the opposition as soundly as possible. It could have worked well. On the other hand, I didn't think I would have necessarily enjoyed driving for Ferrari himself as much as I enjoyed driving for Rob Walker. I felt that driving for a factory team, yet not having the boss around to see how things were going, would be a shame. Being around Rob Walker was part of the fun of racing to me. I decided that driving Ferrari's cars for Rob's team would have been the best of two great worlds.

In the end, the final decision was actually made to do just that. Ferrari was to supply the cars for Rob Walker's team and I would do the driving under Rob's colors. But that same month, I had my crash at Goodwood which put me out of racing and we never had a chance to see how the set-up would have worked out. Yet, had we gone ahead, I'm sure we could have been successful and would have enjoyed our partnership.

Of course, the Ferraris I did drive tended to be sports cars rather than Grand Prix single-seaters. But the Ferrari pedigree was there in full in some ways. Since these were the cars entered for the classic long-distance races, they were the ones which possessed the characteristic Ferrari virtues of strength and endurance in the fullest measure.

I can still remember the first Ferrari I drove—after the dampening Formula Two excursion at Bari, that is. It was a Tipo 290 Mille Miglia, which I borrowed for the Nassau Trophy race on December 8th, 1957, after my own Aston had unfortunately been wrecked by Ruth Levy. I think Ferrari had finally grown disillusioned by the poor performance record of the big four- and six-cylinder cars like the Tipo 159, which the Mercedes SLRs had been leading home in most of the classic sports car races of the middle Fifties. So back he went to the car he has always done best, the classic V-12. This time he used a 73 mm by 69.5 mm configuration of 3.5 liters and fitted with a single overhead camshaft for each cylinder bank, and using the twin-plug cylinder heads derived from the Grand Prix cars of 1951.

This was the 290 MM, and it was a success from the start. In April 1956 it made its first appearance in the Mille Miglia, just about the most brutal baptism which could be devised for any car. But it ran magnificently and Castelotti won at an average of 85.40 mph, followed home by Fangio in fourth place in an identical car. It was more than a year and a half later, when the works team had switched to an enlarged double overhead-cam, wider-bore 3780 cc version called the Tipo 315, that I came to drive the car at Nassau. Yet it was

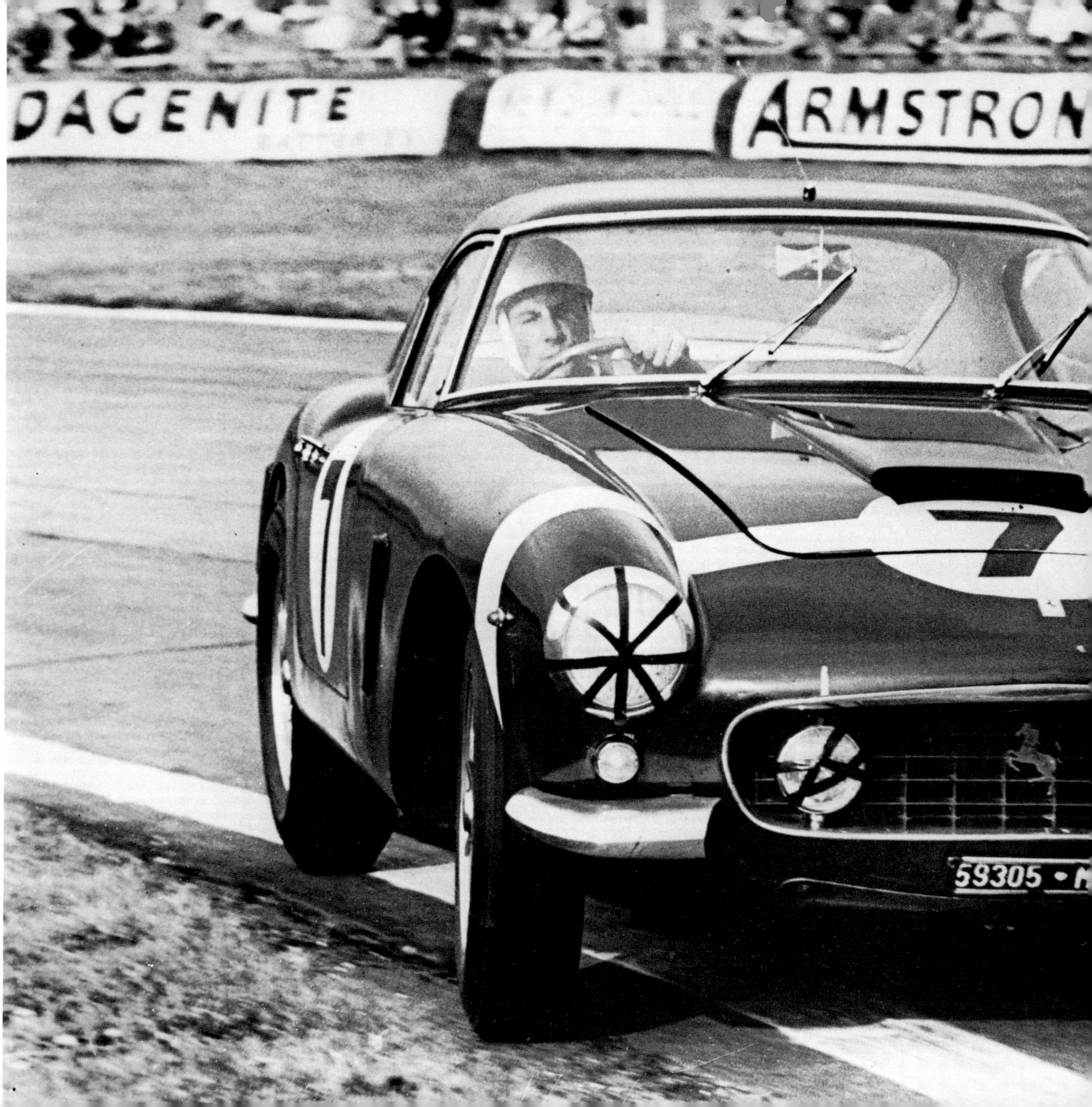
DAGENITE
ARMSTRON
7
59305

On the way to victory at Goodwood.

still a splendid piece of racing machinery, fast enough and reliable enough for me to win, at an average of 101.6 mph.

Earlier in that 1957 season, the new Tipo 315 had been replaced in its turn by the 412 MI, which had a longer-stroke version of the V-12 engine measuring 4.1 liters. Portago crashed one in the Mille Miglia, killing himself and a number of onlookers, which led to the race being discontinued. All the same, Taruffi won the race in a 412 and Collins and Phil Hill won the 1000 Kilometers of Venezuela in a similar car, before the cutting of the capacity limits for the World Sports Car Championship to three liters for the 1958 season put them out of work. But many private owners continued to race them in America, and I drove one in the Havana Grand Prix of 1958.

But the Ferrari I came to drive most often was a hotted-up 250 GT Berlinetta with a three-liter V-12 pushing out more than 300 hp. In its first season this car helped establish Ferrari domination in races like the Targa Florio and Le Mans and brought home the World Championship. Even in 1961, fitted with new bodywork and independent rear suspension, the 250 engine was still capable of winning Le Mans in the hands of Phil Hill and Olivier Gendebien. And it certainly did me proud. Three-liter Ferraris took me to wins in the Tourist Trophy races at Goodwood and Nassau in 1960 and again in 1961, and victories at Brands Hatch and Silverstone. In my final racing season, 1962, we might have done better still: in the three-hour Daytona Continental I won the touring class, coming in fourth overall, two laps behind the leader. And in the Sebring Twelve-Hour race, Innes Ireland and I were doing quite well, with fastest lap to our record when an early refueling stop led to our being disqualified. With a record like that, over five seasons with interruptions, you can't help thinking highly of a car.

During the four or so seasons in which I drove Ferraris, I ran ten races in them. The single breakdown I had was caused by a broken fan blade cutting the water pipe and allowing the engine to boil. Correctly, the fan should always be removed for racing and this was an oversight. In all my other drives with the car, it was fast enough and tough enough to win overall. The only exceptions were that Cuban race, which was stopped after a particularly bad crash, and my being disqualified for refueling too early at Sebring in 1962.

So, whenever I think of Enzo Ferrari, I tend to remember the stamina and speed of the sports cars I drove and, of course, wonder what might have been. It would have been exciting and challenging to work with a man who is such a highly unusual combination of a creative artist and yet a complete professional at the same time. To this day, I remain curious about how our arrangement might have worked out. ✠

CHAPTER 9

Extra-Extraordinary Ferraris

by Karl Ludvigsen

December 1954. The Tipo 118 LM by Scaglietti. Ferrari's first six.

Though he's never owned one, Karl Ludvigsen admits to being hooked on Ferraris ever since he read about the first ones in 1948 and saw them in action–a 4.1-liter America and 2.3-liter Inter berlinetta–at Watkins Glen in 1951. He has been writing about cars since 1954, when his first articles–including one he consulted for the following chapter–appeared in the M.I.T. undergraduate engineering magazine.
Karl Ludvigsen is the author of ten books including two in the AUTOMOBILE Quarterly *Library Series, histories of the Chevrolet Corvette and Opel cars. His book,* The Mercedes-Benz Racing Cars, *was honored with both The Montagu Trophy and The Cugnot Award, and he numbers The Ken W. Purdy Award among other writing honors. A resident of Pelham Manor, New York, Mr. Ludvigsen is the East Coast Editor of* Motor Trend *and a Contributing Editor of* AUTOMOBILE Quarterly.

When Enzo Ferrari wrote his memoirs, he explained the marvelous fecundity of his small factory this way: ''The demands of mass production are contrary to my temperament, for I am mainly interested in promoting new developments. I should like to put something new into my cars every morning—an inclination that terrifies my staff, who hasten to point out that this would be rendering a service to tomorrow's customer, but would certainly bring today's production to a standstill. Were my wishes in this respect to be indulged, there would be no production of standard models at all, but only a succession of prototypes.''

Though I haven't tried to support it statistically, I'm certain that Ferrari has produced more different kinds of cars and engines, in proportion to its size and lifetime, than any other auto company in the

world. The factory is so organized, with a minimum of fixed tooling, as to make it as easy to make one as ten or one hundred cars to a given design. Enzo Ferrari has taken full advantage of this flexibility to satisfy his insatiable curiosity about high-performance cars by building them in bewildering variety.

When Ferrari started making cars after World War II he chose the V-12 engine as his powerplant and has since made it into a visual and auditory trademark, as distinctive as the Bugatti horseshoe radiator or the Mercedes star. The various generations of twelves constitute the main stem of Ferrari evolution, first with single overhead cams, then with twin cams, and now, in both racing and production models, in a flat opposed layout instead of the traditional 60-degree vee. These are the Ferraris, the Barchettas, GTO's, Daytonas and Boxers, that we hear the most about, and quite properly, for they are fabulous cars.

In many respects, however, the lesser-known Ferraris are the most fascinating of all. They came into existence for many different reasons. Some were built to race in a specific event. Some were prototypes, either for Ferrari or for other companies, that for one reason or another traveled no further. Some held promise in their infancy, and were nurtured through adolescence and built in some

Tipo 121 LM.

numbers, but didn't survive to become permanent members of the Ferrari family. And some were pure experiments, created, as are many babies, to see what they would be like. They have only one thing in common: they are all interesting.

THE SPORTS CARS

During Ferrari's first years of car building, through the early 1950's, his sports and Grand Prix cars paralleled each other in design, especially in the engine compartment, and were noted for their V-12 power. The original Colombo-designed V-12 began life at the 1.5-liter size that was right for use as a supercharged GP engine then, and was expanded to two liters, for sports cars and also for Formula Two racing, and further to three liters, by stages, for sports car competition and eventually for production cars.

A new larger V-12 came into use in 1950, the work of designer Lampredi. It found immediate and successful use in both sports and Formula One Ferraris, in the latter through 1951 and in the former through 1954, when a 4.9-liter 375 Plus scored a memorable victory at Le Mans against D-Type Jaguar opposition. But this appeared to be the end of the development line for the big sports V-12, a bulky and heavy car on its 102.4-inch wheelbase. A strong sports car challenge was expected in 1955 from the Mercedes-Benz 300 SLR, which was developed from the then-current Mercedes-Benz Grand Prix car. Working with his usual electrifying speed, Ferrari decided to do something similar for his own 1955 sports car campaign. That "something" turned out to be the only in-line six-cylinder Ferraris that have ever been seen in competition.

Of itself the in-line six was nothing exceptional in 1955. It had been used in recent years by Alfa Romeo, Aston Martin, Talbot and Jaguar for racing, with much success, and of course Ferrari's treasured twelves were only two sixes placed side-by-side. But Ferrari and Lampredi thought bigger than all of those but Talbot, whose 4.5-liter sixes had placed one-two at Le Mans in 1950. They prepared two new in-line sixes, one derived from the 2.5-liter Grand Prix four of 1954 and the other an outgrowth of the three-liter Monza four-cylinder sports car of the same year. Both were raced by the Ferrari factory in 1955, and only in 1955.

Ferrari's fours had first been raced in 1951, as a 2.5-liter and then as a two-liter in the then-prevailing Formula Two category for single-seater cars. When the 2.5-liter Grand Prix Formula One was established in 1954, this engine went back up to its original size, becoming the Tipo 625 F1 measuring 94 x 90 mm for 2498 cc. It delivered 230 bhp at 7000 rpm on a gasoline/alcohol fuel blend. This was the engine from which, during 1954, the first of the Ferrari sports sixes was extrapolated.

Designated the Tipo 118 LM (Le Mans), this smaller of the two

1955 sixes consisted simply of the Tipo 625 F1 four with two more cylinders of the same bore and stroke. This raised the displacement to 3747 cc. By changing the shape of the piston crown, the compression ratio was reduced from 11:1 to 8.75:1, since the sports engine had to run on pump gasoline. The two valves per cylinder were inclined at an angle of 58 degrees between them, and opened by twin overhead camshafts driven by a train of gears at the front of the engine. Their valve gear was unusual, consisting of hairpin springs to close each valve, and a mushroom-type tappet, with its own set of dual coil springs, and in its upper surface a large-diameter roller which contacted the face of the narrow cam lobe.

Dual ignition was a feature of this six, which had the head and water jackets combined in one complex aluminum casting. The steel individual cylinders were screwed into threads surrounding each combustion chamber in a manner resembling aircraft engine design practice. After completion, the head-cylinder assembly was bolted to an aluminum crankcase that extended down to provide side support for the massive aluminum caps of the seven main bearings. It was a dry-sump engine, as is usual with racing Ferraris. The necessary pumps were driven from the gears at the front of the crankcase, which also carried auxiliary drives to the vertically-mounted generator and twin magnetos.

A multi-disc clutch at the engine took the drive to the five-speed gearbox in unit with the rear axle. In all respects the chassis, with its de Dion rear axle and tubular steel frame, was like that of the four-cylinder two-liter Mondial sports car of 1954. Its wheelbase was, however, lengthened by six inches to 94.5 inches, still eight inches shorter than that of the '54 375 Plus, to make room for the two added cylinders. The first Tipo 118 LM roadster was completed in December, 1954, bodied by Scaglietti in the sleek, sharp-nosed style inspired by Enzo Ferrari's son Dino. At least (and probably only) four of these cars were made, developing 280 bhp at 6400 rpm while breathing through three 45 mm Weber carburetors. (For its initial tests the first car had a short-stroke crankshaft giving dimensions of 90 x 78 mm for 2955 cc; this was the experimental Tipo 114 engine of 230 bhp at 6600 rpm.)

Single cars figured in the first two race entries by the 118 LM at Buenos Aires, where Gonzalez was disqualified, and in the non-Championship Giro di Sicilia, which Taruffi won outright after more than ten hours at the wheel. He found the six "very powerful, and.... perfect for the fast bits of the Mille Miglia, provided it stood up." But in that race, in which he drove one of the four 118 LM's,

Maglioli's 121 LM at Le Mans, 1955.

"there were differences of opinion between Ferrari and the tire manufacturers. In this Mille Miglia for some reason our tires did not suit either the cars or the course," continued Taruffi, who kept pace with the flying Mercedes of Moss until his transmission broke. Two of the 118 LM's finished, in third and sixth positions.

Another Ferrari six ran in the Mille Miglia, an even more potent car, without question the fastest machine in the race. This was Maranello's ultimate weapon for Le Mans, getting its first test under fire. Not satisfied with what they had wrought in the 118 LM, Ferrari and Lampredi had made a bigger six, the 121 LM. This was an expansion, by two cylinders, of the 750 Monza three-liter four that had been introduced in 1954. The Monza engine was scaled up in bore (103 x 99 mm) and all other dimensions from the two-liter Mondial. It had similar valve gear but with a wider angle, 85 degrees, between its inclined valves, and the same construction with wet cylinder liners screwed into the head.

For constructional reasons, the 121 LM six was given a bore one millimeter smaller than that of the Monza, so the dimensions were 102 x 90 mm and the displacement 4412 cc. It was a massive, impressive engine, with its angular camboxes, intricate ribbing, the vintage flavor of its head/cylinder assembly bolted to its flat-flanked crankcase, and its bank of 50 mm Weber carburetors. Looks were not its only asset. On an 8.75:1 compression ratio it developed up to 360 bhp at 6000 rpm. This made it the most powerful engine in use in sports car racing in 1966 by the whopping margin of sixty horsepower.

Installed in the same chassis as the 118 LM, the new 4.4-liter six showed spectacular speed on the first leg of the Mille Miglia, but was soon slowed by tire trouble. For Le Mans, two more of the big sixes were readied (doubtless converted from the 118 LM's). One of them, driven by Eugenio Castellotti, led the race on the first lap by a 300-yard distance. He was gradually caught up by Hawthorn's Jaguar and Fangio's Mercedes-Benz, which handled better than the Ferrari but couldn't match its speed on the straight, which was timed at 181.15 mph on the Mulsanne Straight for the best 121 LM, the fastest car at Le Mans that year.

Mike Hawthorn had a close-up look at its capabilities: "The Ferrari's brakes were not as good as ours and their behaviour on corners was not all it might have been; but on acceleration Castellotti just left us both standing, laying incredible long black tracks of molten rubber on the road as he roared away." Before the passage of nine of the twenty-four hours, however, all three 4.4's were retired. They suffered from both drive train troubles and cracks in the intricate cylinder heads that frustrated the proper working of the cooling system. This was a weakness of the design that was never fully rectified.

In fact, Ferrari gave up on the big six soon after Le Mans. One

202

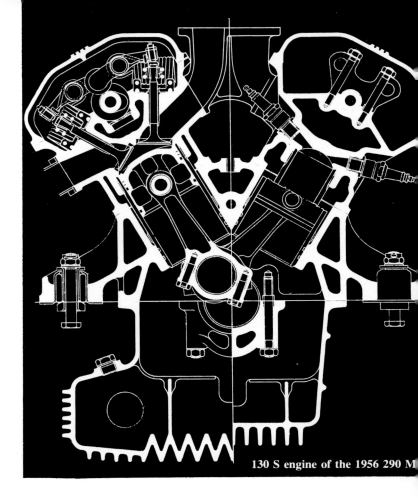

130 S engine of the 1956 290 M

car was taken to a Championship sports car race in Sweden, where it placed third; that was the last works entry of a 121 LM. The factory was reluctant to sell such tricky, potent cars to private owners (as Luigi Chinetti later admitted to a customer) but would-be buyers so pressured them that they yielded. They were rebodied in various styles by Scaglietti and sold to Tony Parravano, Jim Kimberly and Ernie McAfee; a fourth car remained in Italy.

They should not have been sold, because they were terrible cars. The flywheel could only be kept attached to the crankshaft by welding it on. The well-liked Ernie McAfee was killed when he lost control of his at Pebble Beach. "The car was a real poor-handling auto," said "Honest John" Kilborn in a hilarious story about his misadventures at Meadowdale (*Sports Cars Illustrated,* October, 1960) in the

0 S on the dyno.

ex-Kimberly 4.4. "It seemed to be tail-light. It was very unstable in the corners. It spun quicker and easier than any car I've ever had any experience with. The few people I know who have driven one agree with me." By any standards an unsuccessful and unlucky car, the 4.4 nevertheless bemused all who tried it with its staggering acceleration and speed. In both these respects it was the Ferrari counterpart of Bugatti's Type 54.

With a new design team, to which the great Vittorio Jano was an active consultant, Ferrari prepared another completely new sports car engine for the 1956 season. Though it was a V-12, and looked at a glance like the earlier Colombo and Lampredi twelves, it was different from them in every detail. Its camboxes were unusually wide and liberally studded with hold-down nuts, nine along the lower edge

(Colombo engines: six; Lampredi engines: seven). They encased rocker arms with roller cam followers, and inclined valves closed by hairpin springs. Dual ignition was provided, the plugs communicating with the chambers through lengthy flues, and steel cylinders were screwed into the heads. The crankcase was extremely wide. Measuring 73 x 69.5 mm, this twelve displaced 3490 cc and produced 320 bhp at 7300 rpm.

Designated the 130 S, the renascent twelve was installed in the chunky-looking, compact (92.5-inch wheelbase) 290 MM, for Mille Miglia, in which the model scored an outright win in 1956 with Castellotti at the wheel. Later in the season it also won in Sweden and thereby clinched the Sports Car Championship for Ferrari. In '57 the displacement and designation remained the same but the top end of the

203

315 S four cam.

290 MM twelve changed drastically. On the same block, crankcase and front-end drive, it acquired new heads with twin overhead camshafts turned by chains. Made in only miniscule numbers, these were the first-ever four-cam Ferrari twelves for sports cars. During 1957, the only year the factory team used them, they were expanded in size, first to 3780 and then to 4023 cc. In the larger size they were victorious in the Mille Miglia and at Caracas. A limit of three liters on the engine size of prototypes legislated them out of racing in 1958.

In the '58 season Ferrari entered the new V-6 Dino era in Grand Prix competition, giving up the Lancia-based V-8's he had been using and switching entirely to the four-cam 65-degree V-6 engines inspired by his son Dino and designed by Vittorio Jano. The name "Dino" later came to stand for small rear-engined sports and racing cars, but in the first years of this new epoch in Ferrari engineering there were a few front-engined Dinos, too, missing links in the Maranello lineage.

The first V-6-engined sports Ferrari was the Dino 206, which made its racing debut in England in April, 1958. As suited the compact engine, it was a small open two-seater on an 87.4-inch wheelbase, weighing 1500 pounds. With four chain-driven overhead cams, the engine measured 77 x 71 mm for 1985 cc and yielded the excellent output of 225 bhp at 8600 rpm. It drove through a four-speed gearbox attached to the engine to a live rear axle located by radius rods.

This first car ran in two events in April, and was said to be the prototype for a small series of cars like the earlier two-liter Testa Rossa, but this never materialized. A step toward an engine better suited to production was taken in 1959, with the announcement of the 196 Dino. This resembled half of a three-liter V-12 with slight increases in bore and stroke to exactly the same values as the 206 Dino. It was a much simpler engine than the four-cam, having a single overhead cam and rockers on each bank and a 60-degree angle between the banks. This trim little six had a single ignition and triple 42 mm Weber downdraft carburetors.

If Ferrari was trying to attract customers for this car, the 1959 performance of the 196 Dino was not his best advertising. It retired in three out of three major races. A larger 2.4-liter edition of the single-cam V-6 appeared at the Targa Florio in 1960, still in the front of the car but now in an i.r.s. chassis with the disc brakes that had been introduced in 1959; it placed second. These rare single-cam Dinos seemed common at Sebring in 1961, where three were entered. A 2-liter and 2.4-liter were normally bodied, while another 2.4 liter car had a unique body, the shapely new high-tailed form developed by Carlo Chiti for the '61 race cars but without the divided nose inlet that was used on all the others. In fact, this particular Dino, which was raced later in 1961 for Luigi Chinetti's NART by Ricardo Rodriguez, had been used by the factory as a test car for the new body shape.

Rear suspension of a special 1958 Ferrari sports racer.

Other unusual sports-racing Ferraris were created in 1958, the four-cam Dino V-6 in larger sizes playing an important role in these experiments. The relatively simple three-liter V-12 Testa Rossa was chiefly relied on by the team for racing under the new '58 three-liter sports car limit, but a much more elaborate machine was also being built at that time. At a glance it looked like the 206 Dino, but had right-hand drive instead of that car's left-hand steering. And under the skin it was radically different, unlike any other Ferrari sports-racer before or since.

Its front suspension was by coils and wishbones, and its rear wheels were carried by a de Dion tube, which was laterally located by a vertical slide in the back of the differential and sprung by a transverse leaf. The five-speed transmission was in unit with the

differential. In this it was like many earlier Ferraris, such as the in-line sixes, but the design was completely new. Instead of running longitudinally, its shafts were placed transversely, offset to the left just ahead of the differential. In this it resembled the layout used by many Maseratis and Ferrari's own Grand Prix Dinos, but with a further difference: The clutch was fitted into the front face of this transaxle, and in the clutch cover, above the entering shaft from the engine, was the starter motor. (The GP Dino also had its clutch in the transaxle, an inheritance from Lancia, but placed at the left side, to allow an angled drive train, instead of the front.)

An impressive engine, the 312 LM, was the first to be fitted to this chassis. It was a four-overhead-cam V-12 of three liters, with the classic Ferrari dimensions (73 x 58.8 mm), derived from the four-cam

Supercharged sohc 1.5 liter GP engine of 1951.

version of the 130 S used in 1957, as mentioned above. Its output was 320 bhp at 8200 rpm, according to Ferrari. When this car first appeared in competition, however, at Silverstone in early May, 1958, it was powered instead by a big Dino V-6. This was the Tipo 296 of 85 x 87 mm (2962 cc), the largest version of the Ferrari V-6 ever made, rated at 316 bhp at 7800 rpm. It drove the car to a third at Silverstone. The V-6 was then pulled and replaced by the 312 LM V-12 for a race at Spa two weeks later. There it showed speed but did not survive the practice period.

This chassis came to the United States at the end of the 1958 season to compete in the Los Angeles Times Grand Prix at Riverside on October 12. There it was powered by the 412 MI V-12, a four-cam 4023 cc (77 x 72 mm) engine developed from the 1957 sports car engine. This unit had been used in a special single-seater raced at Monza in 1958 (see below) and produced well over 400 horsepower. It served Phil Hill well in a stirring duel with Chuck Daigh's Scarab at Riverside, stopped by fuel pump trouble on the Ferrari. In this form this great and unique car remains in the United States.

In 1961 Ferrari raced his first rear-engined sports cars, and for the '62 season he and Carlo Chiti readied a wider range of such cars. The single-cam-per-bank Dino vee-sixes powered two of the mid-engined roadsters, the 196 SP of 1983 cc and the 286 SP of 2862 cc (90 x 75 mm). They were joined by a completely new addition to the range, a 90-degree V-8. This was the first eight that Ferrari's own staff had built since the war. Vee-eights are rare in the Ferrari pantheon; they include the modified Lancia GP engines he redesigned and raced in 1956 and '57, these sports car engines, the 1.5-liter Grand Prix V-8 of 1964-65, and the production Dino 308 introduced in late 1973.

Introduced as the 248 SP, the 1962 V-8 had cylinder heads like those of the GT V-12's, with chain-driven single overhead camshafts opening the valves through rocker arms. As first introduced it measured 77 x 66 mm for 2458 cc and produced 250 bhp at 7400 rpm; its output was raised to 260 bhp before the season was fully under way by lengthening its stroke to 71 mm, making it the 268 SP (2645 cc). Raced by Maranello only in 1962, the eights were not very successful. Only one 268 SP went to Le Mans, where it retired after being timed at 162 mph on the straight.

Later one of the 268 SP eights came to America, where it was eventually campaigned effectively by Tom O'Brien of New Jersey. In the meantime, Carlo Chiti had left Ferrari, early in 1962, not long after the V-8 was first shown. He joined the ATS organization and designed for them a GT coupe with a 90-degree V-8 engine of 2467 cc (76 x 68 mm), with chain-driven sohc and rocker arms on each cylinder bank—by no means a copy of the 248 SP engine but strikingly close to it in concept. If Ferrari had once had any interest in

The 2.5 liter four of

1950 swing axle GP car.

his own V-8, he certainly would have lost that interest the instant he heard of the plans of Chiti and ATS.

FORMULA CARS

In the world of Grand Prix cars most competitors had used superchargers since that component was proved valuable in the early Twenties. The supercharged era endured through the late 1940's and was ended almost solely by the cars conceived and raced by Ferrari. But when the first Grand Prix Ferraris appeared in September, 1948, they had supercharged 1.5-liter engines, just like the Alfas and Maseratis that were their chief opposition. The use of superchargers

alone ranks them among the most unusual Ferraris ever made and they are discussed in detail in this book's engine chapter.

But as Ferrari phased in his new unsupercharged GP engines during 1950, the blown models were used less and less. The fabulous four-cam machines, which rank among the most intricate cars that Ferrari has ever made, were said at the time to have been sold to the racing Marzotto brothers. One of the chassis, which had rear swing axles and a gearbox attached to the engine, was used for the first race entries of the unsupercharged Lampredi V-12, as a 3.3-liter engine and then as a 4.1. And three of the early single-cam supercharged cars were enlarged in displacement to 1995 cc (60 x 58.8 mm), bringing

207

On these pages, the twin-cylinder GP engine with four valves per cylinder and gear driven cams.

their power to 310 bhp at 7000 rpm. Called the 166 C-America, these cars performed very well in Argentina in early 1951, Gonzalez using one to defeat the 1939 three-liter Mercedes-Benz in its only postwar racing appearance. But Ferrari did not follow up then on his original plan to take those cars to Indianapolis in 1951, coming instead with unsupercharged cars in 1952. And there would be other attempts to compete at Indianapolis as we'll see.

A special four-cam engine was also part of Ferrari's program to stay competitive in Formula Two racing, for unsupercharged two-liter cars, in 1951. Under the 166 F2 designation such a V-12 was built and tested, with more oversquare dimensions (63.5 x 52.5 mm for 1995 cc) than the single-cam V-12 that Ferrari had used hitherto. Its output did not exceed 160 bhp at 7200 rpm, however, which gave this heavier engine no margin over the single-cam V-12, and the new Lampredi in-line four was showing impressive potential, so this four-cam Formula Two twelve was never raced.

Ferrari's four first gained fame as the Tipo 500 F2, which brought Alberto Ascari consecutive World Championships in 1952 and 1953. Then it raced in two-liter form, but it had actually appeared first as a 2.5-liter engine, as part of Ferrari's very effective (and successful) lobbying for a 2.5-liter Grand Prix Formula One to take effect in 1954. The first-ever race entry by a four-cylinder Ferrari was in the non-Championship Bari GP on September 2, 1951, with a 2.5-liter engine, actually 2498 cc (94 x 90 mm). This was bigger in both bore and stroke than the two-liter version of the engine, which measured 90 x 78 mm.

Piero Taruffi placed third at Bari against much bigger cars, a fine first outing for this experimental four. Two weeks later it was used in practice at Monza for the Italian Grand Prix, but not raced. Not until the Modena GP on September 23 did the four appear (victoriously) as a two-liter. With the extra half-liter it was equipped with a pair of the earliest experimental twin-throat Weber carburetors ever seen, with 50 mm throats. With these and an 11:1 compression ratio it produced 200 bhp at 6500 rpm. More engines of this type were used in the F2 car chassis for Formula Libre racing in 1953, paving Ferrari's path to the new Formula One in 1954.

Against the Maserati sixes and Mercedes-Benz and Lancia eights, however, the Ferrari four was able to win only occasionally in 1954 and 1955. Aurelio Lampredi went to work on two new engines which he hoped would tip the power balance. One was an in-line six, the Tipo 115, built at the same time and following the same design principles as the sports sixes described earlier. It was a 2495 cc engine (82.4 x 78 mm) whose output was quoted as 250 bhp at 6500 rpm, not at all bad, but it suffered from some of the same structural problems as the bigger sixes.

The six had been intended for fast tracks, and for slower circuits Lampredi produced an extraordinary concept, the most bizarre Ferrari engine in history: a vertical twin. If the four was good on slow tracks, went the reasoning, a two-cylinder engine might be even better.

The twin's dimensions were 118 x 114 mm for 2493 cc. These were huge numbers by Grand Prix standards but not so large when compared to the cylinders of the unblown Offenhauser engine of that time; in fact the Ferrari twin closely resembles half an Offy, even to its use of four valves per cylinder in a shallow pent-roof combustion chamber. Unlike the Offy, it had two plugs per cylinder, placed in the extreme corners of the chamber. It was built in a three-layer sandwich of aluminum consisting of the head, the water jacket and the deep-sided crankcase, which carried the crankshaft in three main bearings whose caps were cross-bolted for added stability. Each rod journal had double counterbalancing weights, and both possible crankshaft layouts were tried, the side-by-side or 360-degree journal position and the offset or 180-degree layout.

In typical Lampredi fashion, the wet cylinder liners were screwed into the head. A huge breather on each side of the block attempted to cope with the sharp volume changes in the twin's crankcase, another feature reminiscent of the Offenhauser design. A bevel gear at the nose of the crankshaft drove a vertical shaft which rose up to the head, there turning a short gear train to the twin overhead camshafts; part-way up there were takeoffs to the accessories: water pump, fuel pump, oil pumps and ignition. Between the narrow cam lobes and the valves were pivoted fingers with rollers, riding against the cams. Valve springs were coils. Sloping downward to each valve, the inlet and exhaust ports were kept individual, with no siamesing, out to the head faces.

There was no thought of installing this unique engine in the existing Ferrari chassis. Instead, its lightness and small size were to have been exploited by building a much smaller car around it. In the early stages of its development it delivered a maximum of 175 bhp at 4800 rpm, breathing through two special Weber carburetors with close-spaced venturi pairs feeding each cylinder. The need for this engine and car evaporated in 1955, however, when the Lancia D 50 V-8 Grand Prix cars were turned over to Ferrari, and the amazing Tipo 116 was placed quietly on the shelf.

Ferrari did not cease experimenting when the Lancia equipment was entrusted to him. Appreciating that his Super Squalo had appeared to be underpowered, he set aside one of these long-nosed, side-tanked cars and had his engineers fit it with a Jano-designed Lancia D 50 four-cam V-8 engine. This made an impressive-looking hybrid, with a deep scoop down the middle of its louvered hood. As a trial under fire it was taken to Argentina for the first GP of 1956. There this Ferrari-Lancia was driven by Gendebien, then new to Grand Prix racing, who found it a handful but still managed to place fifth in the GP and sixth in a subsequent non-Championship race. The car was not raced again.

Ferrari tried another hybrid far from home, in Florida, at the end of the 1959 Grand Prix season. This was the era of the V-6 Dino, with 2.5-liter four-cam power, and three such Ferraris were sent to Sebring

Maurice Trintignant in the first F2 Dino 156 at Rheims in 1957.

for the first modern U.S. Grand Prix with a new and effective independent rear suspension system. The fourth entry, for Phil Hill, kept de Dion rear suspension but broke with the others in the use of the simpler single-cam Dino V-6 engine. This was the first appearance of the 2.4-liter edition of this engine that was used in a sports car early the following year at the Targa Florio (see above). As tuned for GP competition, it produced 225 bhp at 8000 rpm. Its practice time was the slowest of the four Ferraris, but only by 0.4 second. Painted blue, this one-off car retired with clutch trouble in its only racing appearance.

These are some of the most remarkable and rare Ferrari Formula cars. There are others that led highly abbreviated lives. One was the front-engined Dino Formula 2 car, the first Ferrari to carry the four-cam 65-degree V-6 engine. It first competed at Naples in April, 1957, where it placed third, and in July won the Formula Two race at Rheims. In somewhat changed forms it appeared occasionally in 1958 and '59, and then raced just once, in a quite different configuration, at Syracuse in 1960. There it had side-mounted fuel tanks and all-independent suspension, used effectively by Wolfgang von Trips to win the race. All future 1.5-liter Dinos were rear-engined.

Ferrari's transition to engines behind the driver had begun with another car that ran in just one race, his first mid-engined GP machine. This high-tailed car with Tipo 246 four-cam Dino power was driven at Monaco by Richie Ginther in May, 1960. It was pushed over the line to be awarded a sixth place after stripping its gears. It practiced but did not race at Zandvoort, and did likewise, in a lower configuration, at Monza in September.

Well into the mid-engined era, Ferrari announced a new and advanced GP model for the 1962 season that was shown to the press but never raced in anger. It kept the 120-degree V-6 1.5-liter engine that had been so successful in '61, with one important change: the use of four valves per cylinder. This was then a novel notion in the GP world; only several years later did the four-valve head return to its present dominance. But this engine never went to the track. Used only a few times was that car's other new feature: a transaxle that placed the gearbox between the engine and the final drive, another concept well ahead of its time.

More than a decade later, another prominent Ferrari no-show was the first Tipo 312B3, revealed at the time of the Italian Grand Prix in 1972. An abnormally compact and chunky car, it had a spade-like hammerhead nose punctured by ducts to the two water radiators, and rear bodywork completely enclosing the engine. It gave no satisfaction in testing, Mario Andretti observing cryptically that it would have made a better Indianapolis car than a Grand Prix car. This was a clear indication that its handling was much better on fast turns than on the

Trintignant with the F2 Dino.

The original Dino 156.

many slow ones that wrinkle many modern Grand Prix circuits.

OTHER NOVEL FERRARIS

Interest in Indianapolis, that magnet of money and prestige in North America, weaves a recurring melodic theme at Maranello. When Alberto Ascari drove an unblown 4.5-liter Ferrari there in 1952 he was showing a respectable pace until a wire wheel broke and retired him. Even though three other privately-owned Tipo 375 V-12's failed to qualify that year, Ferrari's budding romance with Indy, fanned by Luigi Chinetti in New York, continued to flame.

The Scuderia Ferrari filed a 1953 entry in the Indianapolis 500, naming Ascari as driver, for a supercharged three-liter car. Built and tested for it was a special engine, the Tipo 250 Indianapolis, which set the Maranello dynamometer quaking as no other engine would until the Tipo 512 of 1970. Built with Lampredi big-block V-12 components, it had the same square dimensions (68 x 68 mm) as the 250 Europa. With a single overhead cam per bank, and a single-stage supercharger, it registered a peak power reading of 510 bhp at 7000 rpm. It never made the transatlantic journey, however.

In 1954, a new purpose-built Ferrari, based on the Tipo 375 4.5-liter Grand Prix car, came to Indianapolis. It was entered in the name of Mrs. Marion A. Chinetti, backed by a group of American Ferrari enthusiasts, nominally sponsored by *Car Life* magazine, and rumored to be driven by Luigi Villoresi. The factory had taken special pains to adapt it to the needs of Indy. The body had prominent side bulges like those of the Squalo, the one on the left housing a supplementary fuel tank, giving the car a left-hand weight bias, and the right-hand bulge containing only an oil radiator and a suitable air scoop. It still had wire wheels, the Achilles heel of the 1952 car at Indy, but in a strengthened form.

Though this car was in the United States early in 1954, it didn't reach the Speedway until just days before the last chance to qualify. Villoresi was not at hand, and none of the drivers who tried this singular Ferrari, including the versatile Freddie Agabashian, could reach qualifying speed. Later this potent car was driven in the U.S., in a hill climb by Carroll Shelby and twice at Daytona, by Bill Holland and Bob Said. It was used as a practice car at Indianapolis in 1956 (see below) and was raced in The 500 Miles of Monza in 1958 by Harry Schell, without success. To meet the rules of that event its displacement was reduced to 4.1 liters. Thereafter the car was rebodied, after the style of the 1960 GP Ferraris, and returned to the United States, where it was acquired by Dieter Holterbosch.

What seemed to be the most promising Ferrari Indy challenge of all was mounted in 1956. It traced its beginnings to early 1955, when Dr. Giuseppe ''Nino'' Farina sought to satisfy a craving to race at Indianapolis. Then forty-six years old, a former World Champion (1950) driver, the Turin-born Farina wanted to round out his long racing career with an Indy 500 appearance. He arranged backing for this from the Bardahl Oil Company through its Italian office in Florence, and a Ferrari-powered car was made ready.

Farina and Bardahl sidestepped the problem of building a special chassis for Indy by buying a brand-new 500D car, less engine, from Frank Kurtis. This was shipped by air to Italy and delivered to the Ferrari factory, where one of the then-new in-line sixes was installed. On paper this engine, the big 121 LM that was to make its bow in the Mille Miglia, looked just right for the job. It fitted inside the Speedway's 4.5-liter limit with its 4412 cc displacement. With its six big cylinders it promised good torque, for acceleration off the turns, plus an ability to rev high to get straightaway speed. Its installation in the Kurtis required the casting of a special transmission case and the making of new gears, and this unexpected work absorbed the time that had been set aside for testing the Bardahl-Ferrari before the 500. Bardahl decided against shipping a completely untried car to the Speedway and it remained in Italy through the 1955 season.

Some special development work was done on the 121 LM engine for its Indy appearance, but much of it was inconclusive. Officially its output was 380 bhp at 6500 rpm, up 30 bhp and 500 rpm from the rating of the Le Mans 4.4 engines. This was achieved with Weber carburetors and a higher compression ratio on gasoline-based fuel. Hilborn constant-flow fuel injection was also tried in Italy, but the testers substituted their own pressure pump for the Hilborn pump, which to their eyes looked too inefficient. This may have been one of the reasons why they registered forty horsepower less with the injection than they did with carburetors.

Final preparation of the car for the 1956 Indy race was entrusted to the O.S.C.A. shop of the Maserati brothers in Bologna. There it was completely dismantled, measured, checked and reassembled. It was a handsome car as it was readied in Italy, with its Ferrari-like open nose scoop and its twin exhaust pipes running high along the left side of the cockpit. In this trim it was tested by Farina at the banked concrete Monza oval on March 31st, 1956. It did not arrive at the Indianapolis track until May 12, however, shortly before qualifying. Farina had preceded it, using the special V-12 Ferrari Indy car to complete the prescribed forty laps of his driver's test in only forty-six total laps, and impressing and pleasing the Speedway denizens with his modest and friendly attitude.

In and around the Bardahl-Ferrari crew the atmosphere became less cordial when it was discovered that the car wasn't able to reach racing speed. The slowest qualifier of the twenty-nine cars that made the field on the first weekend bettered 140 mph, while the best Farina could manage his first times around was 131. Not yet used to the special technique needed to go fast on the four-cornered track, he was braking so hard that the discs were turning blue. Set up too tightly in Italy, the rear axle was smoking after five laps.

There were sharp differences of opinion among those tending the car about the proper fuel to use. Farina and the Italian contingent were reluctant to give up the gasoline-based fuel on which the engine had

214

A pair of Ferrari Indy cars, both unsuccessful. Below: The 1952 Ascari car Above and opposite: Nino Farina's Bardahl-Ferrari built on 500 D Kurtis chassis for the 1956 race.

been weaned, while those who knew Indy were certain methanol was the answer. The engine was overhauled; Hilborn injection was fitted; and a new low-level exhaust system was made with a single collector pipe. A much lower gear ratio was fitted too. The best speed with all these changes was 136 mph on methanol. Frustratingly, the six refused to rev above 6100 rpm, even with the very low axle; suspected were the relatively heavy valves and valve gear (compared to the Offy's four per cylinder) and turbulence in the sharply tapered inlet ports. Though a dose of nitro might have boosted the lap speed to 140, even this wasn't fast enough to assure a place among the thirty-three starters, so the project was abandoned.

Two years later Ferrari built two special cars for a 500-mile race that was held so close to Modena that he could hardly ignore it. He was further encouraged to enter something in the 500 Miles of Monza by the inclusion of that amazing race by the Automobile Club of Italy in the list of events that counted for its annual prize, carrying a big cash bonus, to the most successful Italian racing car maker. And Ferrari's men did their work well. One of his two cars was the fastest qualifier and third-place finisher among a strong field of Indy cars in the race, on June 29, 1958, which was then the fastest ever run.

That fast Ferrari, a car that amazed the Indy delegation that came to Monza, was a fascinating hybrid used only for this race. It was built around one of the four-cam 1957 sports car engines at maximum displacement (4023 cc, 77 x 72 mm), dubbed the 412 MI—for Monza-Indianapolis. With six twin-throat 42 mm Webers it delivered a very respectable 447 bhp at 7700 rpm. It was installed in a new chassis similar to the 1951 Grand Prix Ferrari with more frame bracing, coil spring front suspension, and only three forward speeds in the rear-mounted gearbox. Its smooth if tall body had a high rear headrest and a deep wrap-around windshield.

With a heart-stopping display of sheer courage on the bumpy Monza bankings, Luigi Musso qualified at 174.653 mph to take the pole in this magnificent Ferrari. He led for a few laps in the first of three heats, then was relieved by Mike Hawthorn, placing sixth. Musso started the second heat and was relieved again, this time by Phil Hill, who later called his stint at the wheel of this bucking brute ''the toughest job I ever tackled.'' Phil took over from Mike Hawthorn in the third heat and ended up with more laps in the car than either of the others, all of them adding up to a third-place finish. The car had done its job; it was never raced again. Its engine was pulled and installed at the end of the 1958 season in the special sports car that was sent to the U.S. for Phil Hill to drive at Riverside (see above).

The other Ferrari for Monza was less ambitious and less successful. It was a low single-seater that closely resembled the 1958 GP cars from Maranello. Ferrari had begun work on it early in 1958, thinking both of the Monza race and also of another Indy challenge, for which

Dino 296 V-6 for the 1958 Monza 500.

he felt a light car, easy on fuel, might have the legs of the big American roadsters—a view in which he was just a few years ahead of Colin Chapman's Lotus-Ford. He used the chassis of the original 1957 Formula Two Dino, replacing its transverse leaf rear spring with long coil springs and moving its oil tank from the tail to the left side of the cowl.

Under the hood of this car's clean, rounded body was a 296 MI engine, a 2962 cc four-cam Dino V-6 of 316 horsepower. In fact it was the same engine that had powered an experimental sports car to third place at Silverstone more than a month before (see above). The V-6 Ferrari was qualified for the 500 by Phil Hill at 161.226 mph, toward the back of the field, and survived only eleven laps of the first 63-lap heat. An excess of oil on the ground betrayed a major engine failure, though the official cause of retirement was magneto trouble. Whatever the reason, it dimmed Ferrari's hopes for a performance good enough to interest some of the visiting Americans in buying such

a car.

About a year after this stormy Monza happening the Ferrari staff was hard at work on a new prototype that could not possibly have been more different from the big red rocket that it had taken three men to tame. Enzo Ferrari had begun a project that he hoped would help him in several ways. He wanted it to answer those critics who said he could only build big, powerful cars and engines. He also hoped it might show that his hard-won racing know-how could make a direct contribution to better road cars. And he thought it might make some money for his firm. The project was a new small engine and a car around it.

At the end of 1959 a new sound was radiating from one of the Maranello dynamometers, that of a tiny four-cylinder engine, the Tipo 854. It displaced only 849 cc (65 x 64 mm) and had a chain-driven single overhead cam operating inclined valves through rocker arms in the classic Ferrari manner. In this respect it was very advanced, even

The 412 MI engine of a 1958 Monza 500 car.

exotic, for its day; but this is a layout which, with crossflow porting, is now virtually standard on the best small-car engines in the world. Ferrari developed the all-aluminum 854 in several states of tune: 64 bhp at 6000 rpm, 72 bhp at 7000 for a GT model, and 84 bhp at 7000 for a sports edition.

In 1960 Ferrari was quoting 75 bhp at 6800 rpm for the Tipo 854 engine fitted in a prototype coupe, bodied by Pininfarina, that was being demonstrated to friends and to firms that might be interested in building such a car. It carried a star and a tommy gun as its identifying emblems, and rolled on a 86.6-inch wheelbase and 13-inch wheels. There were no immediate buyers for the concept, however, and little was heard of the ''Ferrarina'' until it appeared in a completely new form at the Turin show at the end of October, 1961. There Bertone exhibited a fastback coupe called simply Mille, for the over-1000 cc capacity of its engine. Changed in detail from the Tipo 854, it had square dimensions of 69 x 69 mm for 1032 cc and generated 95 bhp at 7000 rpm from that tiny package. Maranello breeding was self-evident in its tubular frame, coil and wishbone front suspension and rod-located leaf-spring rear axle, though the Ferrari name appeared nowhere, and never would on this car.

Twelve months later the Mille had found a home. Its manufacture was to be taken up by the de Nora family of Milan under the new name of ASA, which showed the ASA 1000 GT at Turin in 1962— identical to the Bertone-bodied Mille. Another year passed before the actual production design was frozen, and a year more before production was actually under way. ASA ventured into racing, built in 1966 a car called the Roll-Bar which bears a striking resemblance to the 1968 Chevrolet Corvette, and introduced a 1755 cc version of the four (69 x 90 mm) and a remarkable small six, too, developing 124 bhp from only 1290 cc. In spite of (or because of) this creativity, ASA was no longer among the world's car makers in 1969.

In the meantime Ferrari had arranged for one of his designs to be built by a rather larger firm, and this was unveiled over the 1966-67 winter as the front-engined V-6 prototype that remained stillborn. Much like the ASA, it came into being with an engine by Ferrari and a body, in 2+2 coupe form, by Bertone. It was built in 1964 for Innocenti, which was then making Austins in Milan under license.

Ferrari's contribution to the Innocenti Coupe was a variation on the single-cam Dino V-6 theme he had initiated with the 196 Dino in 1959 (see above). Many had suspected that this engine had been conceived with volume production in mind, and its use in the proposed Innocenti model was proof that those suspicions were well-founded. As supplied to Innocenti, the 60-degree V-6 was of 1787 cc (77 x 64 mm), and produced about 120 bhp at 6200 rpm. The front-engined Coupe's chassis and suspension were of conventional design for its day. After limited testing, Innocenti decided to shelve its plan to make a 2000-car series, remaining true to its principles of large-volume car production. The prototype was preserved, however, and is still owned by Leyland-Innocenti.

Such is our survey of the remarkable Ferraris, the unique and experimental models that Maranello has built with such astonishing facility. Each was of vital and immediate interest to Enzo Ferrari while it was being made. But once it had served its purpose, or proved or failed to prove its point, Ferrari dismissed it from his mind. Few of these cars and engines held any sentimental attachments for him. Once an engine had been built, tested and raced, it had revealed its secrets to him. He could store, sell or scrap it without a moment's regret. To do otherwise would have been to clutter his small shops with cobwebbed memories. It is for others to save, polish and revere the machines created by Enzo Ferrari. It is for him to devote, as he has, a lifetime to the art and science of fast cars. ⚜

10 A Championship Season and Other Memories

by Phil Hill

Phil Hill, America's only World Champion driver, needs no introduction. Since his retirement from racing, he has pursued his many diverse interests including classical music and antiques, and turned to business activities. Today he is deeply involved in the restoration of antique automobiles. His own Packards have appeared in AUTOMOBILE Quarterly *as has his account of what it's like to drive a Mercedes-Benz SSK. Mr. Hill would agree that it is difficult to make reliable generalizations about his former employer, Enzo Ferrari. Too much depends upon the outlook of the particular driver in question. But Phil Hill's experiences were not atypical. In this chapter, Mr. Hill describes—perhaps more candidly than ever before—his impressions of Enzo Ferrari and the events which led to his ultimately quitting Ferrari's team.*

Looking back on it, I think my first serious meeting with Enzo Ferrari set the tone of our relationship for years to come. It was an awkward meeting for me, and he must have sensed how I felt and reflected it. That awkwardness was to remain between us until very recently.

It was early summer of 1955. I was twenty-eight years old. Behind me were publicized successes racing Ferraris in California and a second place in the latest Mexican road race, and now I was in Europe along with Richie Ginther, who had ridden with me in Mexico. There had been a tentative offer from Briggs Cunningham to drive for him at Le Mans and a possibility of campaigning Allen Guiberson's Sebring Ferrari with Richie as my mechanic, but now I was on my way to see Enzo Ferrari in Modena at his summons. I had met Ferrari before

Running hard at the Dutch GP in 1961.

under more casual circumstances, but this was different—a command appearance. The romance of it all was not lost on me on that great train ride down the Riviera and into Italy.

The factory was in Maranello, a few miles into the Apennine foothills from Modena, and I still see clearly those dark offices where I waited and waited for my audience. In later years I saw many instances of Ferrari's mastery in putting even the most important visitors off balance with long heel-coolings in those dim chambers. And there I was—inexperienced, expectant and becoming increasingly unimportant by the minute. When I was finally ushered in I was made even more ill at ease by Ferrari's patrician, nose-in-the-air manner. At the peak of my discomfort he suddenly just came right out with it: "How would you like to drive for me at Le Mans?"

I suppose the expected answer, after all the tension of my waiting and his posturing, was a grateful "yes" in an outburst of relief. But my enthusiasm was tempered by a vague sense of responsibility elsewhere—the Cunningham offer, Guiberson—so I was also hesitant. Clearly he was not pleased by my equivocation, and my tension increased even more. In later years, I learned that Ferrari was satisfied with no less than one hundred percent enthusiasm from his drivers at all times, no matter what the circumstances.

Then he started talking about the Mexican road race in which I had finished second to Maglioli. (Ed. note: Hill's three-year-old 4.5-liter Ferrari was some 20 mph slower than Maglioli's 4.9.) "What do you think of our great protagonist?" he asked, meaning Maglioli. I answered, "Oh, I think he's just fine."

"Well, I'm going to put you both in the same car. What do you think of that?"

"Oh. Oh that would be great. That would be fine. I look forward to it."

"Well," he said, "come now and let me show you the great cars you will be driving," and conducted me to the factory. His remark struck me as pompous and did nothing to decrease my discomfort. But, of course, I gladly went with him to look at the cars. He proudly indicated a six-cylinder engine on the dynamometer. I had been driving twelves, and loved them, and here he was asking me to admire what to me was a commonplace in-line six with overhead camshafts. I had raced Jaguars. There was nothing here new to me. I was baffled by the fuss he was making.

"What do you think of it?" He was like a new father expecting compliments for a squalling prune. There I stood in that immaculate factory fumbling for an appropriate response.

"Very nice, Commendatore."

"Don't you think it is beautifully symmetrical?"

"Yes," I said weakly. "Yes, it is."

Test sessions at Modena during 1959.

My uneasiness did not slacken any after I was actually on the team. The experience was all so totally new to me and I had no frame of reference in which to judge things. I was breaking new ground and there was no one to warn me of the chuck holes. (In reaction to this I swamped Dan Gurney and Richie Ginther with detailed briefings when they later appeared on the scene, probably to their infinite mystification because, of course, I could tell them only what I wish *I* had known.) Actually I was a victim of my own myths. I naïvely believed everyone at Ferrari to be blessed with total wisdom— all-knowing, all-seeing, all-understanding—so that all I had to do to be privy to the secret was to follow their instructions. From the start they continually told me to take it easy—*"piano, piano"*—and I got the idea that to damage the car in any way would be the worst disaster on earth. So I did, indeed, take it easy. But then I could see that they still were not pleased. Here I was doing exactly as they told me and yet disappointing them. I was distraught.

This confusion kept me from being wholly instinctive about my driving at Le Mans that year (1955). Even though I had the third fastest time—our car was a six-cylinder 4.4-liter—I was painfully careful all the time. It was a terrible strain. We were in third place when we went out because a failure in the engine casting had resulted in a water jacket leak. That was after the awful Levegh crash and there was an understandable pall over everything anyhow.

Maybe the car's breakdown was a proper initiation to the team for me because it gave me a chance to see how Ferrari reacted to adversity. He raised holy hell when something broke, storming at the engineers demanding to know exactly why. "Why didn't we find this out in our tests? We run these engines on the dynos all day. We take them to Monza and run them. Why did this happen?" And the engineers would say, "Ah, Commendatore, there's something about going down that long straight at Le Mans that is different." He always demanded an explanation from the engineers and I came to find out that as often as not, their explanation was to blame the driver.

By 1961, the year I was to win the World Championship in Formula One as well as winning Sebring and Le Mans for the second time, I understood a lot better what was really expected of me. I understood now that although they said *"piano, piano"* they meant let it all hang out. I also, literally, understood more that was said around me. When I first started with Ferrari I managed by communicating in French, but then one glorious year I was suddenly aware that I was understanding Italian perfectly, even the Modenese dialect. Needless to say, I didn't let on at first. Whether my open eavesdropping really made any real difference or not, it at least gave me a sense of being one-up, a confidence-builder that I needed. And I was even beginning to feel at home with all the loony unpredictability, that *buffo* atmosphere that everyone who has ever driven for Ferrari has a whole evening of stories about. Peter Ustinov in his "Grand Prix of Gibraltar" captures the feel of it brilliantly.

Maybe I would have found a more rational milieu at Lotus, for instance, but I preferred to stay with Ferrari. It might have been a preference for the devil known over the devil unknown, but there were other reasons. Ferrari was the pinnacle, and I enjoyed being a part of that. And, equally important, I had—to my relief—discovered that all was, after all, a far cry from perfection at Ferrari. So I didn't have to worry about being perfect myself. When you come to see that the emperor has no clothes you are more comfortable with your own nakedness.

I must digress for a moment to say that looking back on that championship year I realize that there was more on my mind than Ferrari and the Ferrari team. In retrospect, I would say that my state of mind was much like Mike Hawthorn's must have been in his championship year of 1958. There was a conflict. In short, I was wondering what the hell I was doing as a racing driver. Though I could not have articulated it then, my total being was not in harmony during 1961. I had been to many races and to many funerals and a battle was mounting within me. There was an inner drive to race, to excel; but there was also a tremendous desire to stay alive and in one piece. The only time I was totally free of that conflict was in my early sports car years.

I wish I had begun my Formula One career the first year I drove sports cars for Ferrari. I would have been able to do my best job for him then. But they would have said: "Oh, you're too immature to drive Formula One," or some such drivel. They built up such a mystique about Formula One, as if one must serve a novitiate for it. Nonsense. My skills were as ready for Formula One in 1955 as they ever were. All one had to do to disperse the mythic aura around Formula One was to drive Formula One. An opportunity, incidentally, Ferrari almost teasingly withheld from me Grand Prix after Grand Prix in spite of my successes in sports cars. I finally forced the issue by borrowing a Maserati from Joachim Bonnier for Rheims in 1958, the week after I had won Le Mans the first time for Ferrari and still he had stalled saying: "*Aspettiamo, vediamo.* Wait and see."

It is too bad. Ferrari might have had a better part of me earlier. Not that I was slower, it is just that the inner conflict was a drain on my energies and I was not of a single mind about racing.

Some drivers seem never to entertain that conflict. They go right through their entire careers without a hint of it. I think Wolfgang von Trips was one of them. He was tremendously turned on by everything about racing. The driving, the adoration. His inner image of himself seemed to be as a racing hero. That was his completion. But maybe

These pages show Phil during Monza test runs and describing his impressions (below) after an especially hot session.

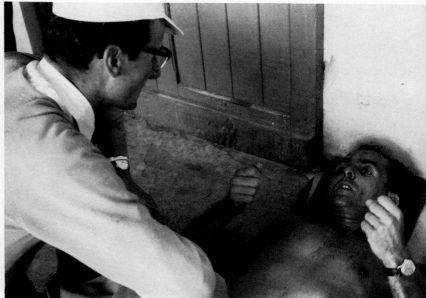

Two 1959 events: Zandvoort (above) and Monaco (below).

the ambivalence had not yet touched him, or maybe he was particularly adept at concealing his feelings. Now, especially, as I look back, I find it hard to believe that anyone doing something as irrational as gambling so openly with his life can be totally free of conflict. What you do, of course, is either quit—walk away forever—or accept the risk as if it were a contract and then forget it. The danger comes, I think, in taking caution with you in the car, chewing at your instinctive responses. Doubt must be left in the pits along with your spare goggles and street shoes or it can kill you.

I want to emphasize that my awareness of the dangers of racing sprang from a sense of my own mortality and was not connected in any way to the cars I drove for Ferrari. The race cars were designed with a sense of structural integrity that was never violated—no drilling of the chassis for lightness or risky shaving of any safety margins. And the cars were built by a handful of devoted men, the holy of holies at the factory, who took great pride in their craftsmanship. With more than any other marque I can think of, it was simply unnecessary to be concerned about the chassis breaking. As Luigi Chinetti used to say: ''With Ferrari you have not to worry. You get in, you drive, you win.'' He was right, at least, about not having to worry.

The car for the 1961 season complied with the new formula, a drop from 2.5 liters to 1.5 in engine size. It was the V-6 Model 156, largely the work of Chief Engineer Chiti. Chiti was a big pudgy man with a soft round face and round glasses who had an artist's temperament that could throw him into almost childish tantrums when his wishes weren't immediately satisfied. He was capable, inventive, and an absolute workhorse, and I liked him.

Everyone had been grousing that the new 1.5 liter formula was going to be gutless and dull, and we drivers were expecting the worst. Maybe that is why the 156 felt so good. It had a five-speed gearbox, nonsynchronized, that was astoundingly easy to shift. Nothing to it. Ferrari could make beautiful gearboxes, contrary to later reputation. The only difficult Ferrari gearboxes were the synchronized ones used in the later Gran Turismo cars. These were heavy to shift, had too much travel and balked when you tried to hurry, but there was never a problem with the Formula cars.

The 156 was smooth to drive and had an advantage of some 30 hp over cars from other factories that year, but in some spots I would have happily swapped that advantage for the agility of, say, Stirling Moss' Lotus at Monaco. Not that the 156 was clumsy, but it was nowhere near as nimble as the Lotus when it came to handling. When this became apparent, we were instantly in the midst of another classic pit brouhaha. To improve handling we had to go to lots of negative camber in the rear wheels and that gave the tire people fits over the incredibly intense heat build-up on the inside edge of the tire.

"They won't last the race," they'd tell us. There was much waving of arms and throwing up of hands, with spectators shaking their heads at our knock-kneed cars. "Look what those Italians had to do to make those things handle."

What bothered me most about the new car was something that had bothered me about previous Ferraris—cockpit heat. And this matter, as simple as it seems, was really at the core of what it was like to drive for Ferrari in those years. To appreciate it requires a little understanding of the Italian social system. There seems to be an acceptance there of being whatever you are for life—a waiter, a mechanic, a doctor—and a fierce pride in doing it well. There is as much satisfaction in being a good hotel porter as being a good tailor—or at least there was then. Which was great in many ways, of course. But there were problems as well, because along with this pride came a jealousy over jurisdiction. And a driver was not to encroach on what was considered a mechanic's territory.

Avoiding stepping on sensitive toes took some delicate stepping indeed, and at first I was tromping in all the wrong places. For instance, when they first had me testing cars at the Modena track I would come in and tell the mechanics exactly what was wrong and what I thought they might do to correct it, and then watch with astonished frustration as they fussed and fiddled with everything else before finally coming with a great "aha" to what I had told them twenty minutes earlier. I chewed a lot of Gelusil before I caught on that the game was not to tell them what I *knew* the trouble to be, but rather to describe around it in an elaborate charade and let *them* do the discovery. Trips had no such trouble. He was completely unmechanical. They liked that.

Clearly such jurisdictional disputes did not make for close mechanic-driver working relationships such as Moss had with Alf Francis or Jimmy Clark had with Colin Chapman and his mechanics. Our mechanics treated our complaint about cockpit heat as if it were an irrational quirk of the driver class. For years the classic Ferrari "solution" to cockpit heat was to duct in more cooling air rather than block the air superheated by the radiator from entering the cockpit in the first place. They could not believe that even the smallest hole could turn our feet to rare roastbeef in a few laps. Then somehow I got the message across to my mechanic, Dino, and he plugged up the holes for me. What a difference it made! But what a lot of energy expended to get such a simple correction. And I somehow felt that, to do it, Dino had stepped outside the accepted mechanic/driver pattern on the team and done it for me personally. Fortunately, by 1961 that easier relationship was more common.

As for Ferrari himself, the drivers rarely saw him. Usually at the start of the season there would be a message that Mr. Ferrari wanted

Rheims, 1959.

225

you to stop by and say hello. You would go, wait the standard half
hour or so, and then be shown in for a light bantering exchange.
"Come stai. How are you? How's your love life?'', that sort of thing.
And Ferrari's language could be x-ratedly blunt. Maybe it was the
incongruity of that Roman-coin face uttering the one-of-the-boys words
that surprised me. Then he might ask what we thought of the racing
program for the year; had we seen the cars? Just small talk. Then we
might go out to dinner. There were a number of excellent restaurants
in Modena, one of which Ferrari actually owned.

I cannot yet see clearly what Enzo Ferrari thought of the drivers. I
do not know if he had any genuine feelings for us as individuals or
whether we were just tools. He was definitely on what would now be
called an ego trip, so that the drivers, even the engineers, were
probably tolerated as necessary evils in his grand scheme of things. He
was like a sculptor resenting the stone that gets in the way of his
glorious vision. In a film script Denise McCluggage did based on a
Ferrari-type character, she has someone say to him: "To you a driver
is just another spare part that unfortunately can't be produced in your
factory," and that strikes me as valid. I suppose wondering what
Ferrari really thought of us is rather like wondering whether General
Patton liked his soldiers.

Ferrari never went to the races, but he was sometimes in the pits
on practice days, especially at Monza which was close to Modena.
Whenever he was present, the atmosphere changed palpably. Everyone
acted differently, and I soon found out why at Monza. The windstream
had been striking me in such a way my vision was blurred. I came
into the pits and stomped around, grousing. In my usual hyperbole, I
complained to the engineers that the windshield on my car was so
poorly designed the wind was rattling my head and I could hardly
even see. Ferrari, who overheard this, wheeled on me and snapped:
"Put your foot down and forget your head!" I was crushed. And I
was properly circumspect in his presence thereafter.

Working for Ferrari could not have been easy for anyone. All the
people at the factory seemed to walk on tiptoe around him expecting a
flash of his unpredictable temper at any moment, and sometimes he
acted with what I considered premeditated cruelty. Tavoni, our team
manager, must have borne more than his share of Ferrari's prickly
disposition; he had been Ferrari's secretary from the time he was
scarcely more than a boy. He was a sensitive person and I respected
him and considered him a friend. One year he said to Ferrari some
two months in advance: *"Prego,* Commendatore, I would like to
take Christmas Eve off to be with my family. Would this be
possible?'' Ferrari thought over this unprecedented request carefully
before he agreed. But come Christmas Eve Tavoni was still working
away—five o'clock, six o'clock—anxiously waiting for Ferrari to

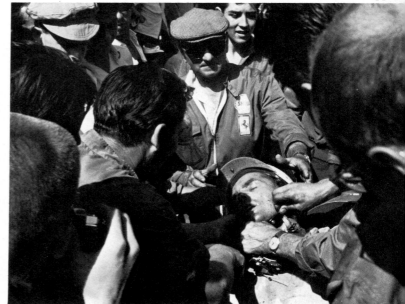

**Formula Two at Rheims in 1959 (above). Heat prostration
after the European GP in 1959 (below). Ginther and Chiti (opposit**

appear and dismiss him. Finally, at 8:30, Ferrari came in and told him he would be needed for another hour. There were many others who had similar Scroogian tales to relate about Ferrari.

And the parallel is not lightly drawn. For Ferrari, as for Scrooge, his work was his life and one day was much like any other. Ferrari's marriage clearly held little pleasure for him, and the early death of his only son, Dino—who was a fine promising young man—that left another hollow in his life. He filled it with work. It was that single-minded devotion to work that enabled Ferrari to accomplish what he did. And he did not seem capable of understanding that not everyone had the same intense focus.

Ferrari's expectations of performance exerted a strong force that radiated throughout the organization, and the drivers were not exempt from it. Indeed, since we were at the most public end of the chain, we felt it more strongly, "heroes" or not. Maybe it is no different on any team—the pressure to do well on the field of endeavor—but somehow there was an added element at Ferrari. Rather than the race being a culmination of a team effort to win, there was a feeling instead as if you, the driver, had been reluctantly entrusted with this gem of a machine, this fruit of genius, and hopefully your natural dunderheadedness would not destroy it.

When one of us did win I sensed a certain reluctance on Ferrari's part to share the laurels with the driver, to pat him on the back and thank him for a job well done. It was more like Ferrari felt the victory was doubly his—he had not only managed to build a car that was better than all the other cars but a car that was also good enough to foil even his driver's natural destructiveness. I felt he expected me to say: "Thank you, Commendatore, for providing me with such a wonderful machine." But I would not do that. Actually, I have always believed that the car was clearly the most important element in winning—maybe as high as ninety percent car and ten percent driver—yet, because I felt Ferrari's reluctance to admit to even that ten percent, I was stubborn in acknowledging his ninety percent. I suppose that was why I would find myself regaling the press with stories and appropriate gestures about the faults of the car—its horrible understeer, the dreadful way it bottomed at the back—leaving the impression my victory had been hard won. True as this might have been, it was hardly politic, and my verbiage did me no good with the organization.

But whatever the difficulties with Ferrari, the 1961 season began with hope. I felt the car to be truly competitive and that a winning season seemed within our reach. But that very possibility of success brought into focus another source of tension peculiar to driving for Ferrari. Instead of being a team, we were a collection of individuals who happened to operate out of the same pit. It was the custom on

Belgian GP, June 1960.

other teams to designate a number one driver, a number two, etc. This practice eliminated intrateam competition and saved wear and tear on machines as well as psyches. Ferrari did not do that. Maybe if a driver was well ahead on points midway through the season, then the other drivers were expected to defer to him on the course. I played the after-you-Alfonse game with Mike Hawthorn both at Monza and at Morocco, pulling up to finish behind him and allow him to collect the final points to give him the championship. It was expected of me, but Mike's gratitude was effusive nonetheless.

At the start of the 1961 season then there was this awareness that we were not only racing the other teams, but we were racing each other to win a designation as Numero Uno from Ferrari, a promised decision which he nonetheless withheld race after race and, indeed, never did make. The tension was excruciating and could not be relieved by a frank expression of competitiveness, not acceptably anyhow, between friends and teammates.

At the opening race at Monaco I finished third after minor problems with carburetion and brakes. Trips was fourth. Then at

Zandvoort Trips got in front and I never could catch him; we finished that way, nose to tail, first and second. At Spa it was just the other way around. Going into the French Grand Prix I was ahead nineteen points to eighteen.

At Rheims there was an incident in practice that points up what childish responses this sort of competition can elicit. I was noticeably faster than Trips in practice, which gave me a guilty surge of pleasure. Then I noticed he was complaining to Tavoni, probably about his car.

Now at the start of the season we had agreed that we were not going to ask to drive a teammate's car, a previously common practice for the purpose of comparisons. But under the pressures of the 1961 season there had begun a creeping paranoia that anyone who drove another's car might abuse it. Yet here was Trips begging Tavoni to suspend the rule—not so he could drive my car, but to have me drive his! Obviously, he believed his car to be considerably slower than mine and my driving it would show that. I didn't really want to swap. After all, it might *not* be running right and they might fix it so that it would be faster than mine. I wanted to hang on to any advantage I could. This was the fourth race, midway through the season, and we all thought that surely Ferrari would designate a Numero Uno after this race.

At that moment a driver came into the next pit saying that there was oil all over Thillois. I thought: "Ah, now's the time to try Trips' car with a built-in excuse for a slow time." His car seemed to be slightly down on power from mine, but not much. The Rheims course is essentially a triangle of fast straights, so any power difference is immediately magnified. There wasn't enough difference to show up that way. And when I discovered that there was no oil to speak of at Thillois after all, I really let it all hang out. I flew.

When I came into the pits, Trips was the picture of gloom. I tried to keep from breathing hard (those three laps had been as close to ten tenths as I could make them) and said: "I'm sorry I wasn't able to turn on a good one with all that damn oil all over, but it doesn't feel half bad to tell you the truth." Well, I had come within a hair of my time in my car and was almost a full second faster than Trips. He was in the depths of despair. I won the 100 bottles of champagne they give at Rheims for the fastest lap in practice.

But hubris will out. The day of the race the sun softened the recently-surfaced course and freed gravel until it was like driving on a spill of ball bearings. Trips went out early with his radiator holed from the flying stones. I was leading and all I had to do was stay pointed in the right direction. Then near the end of the race I spun at Thillois and Moss, who also lost it, clouted me and my engine stalled. The starter wouldn't work. I tried to push start it. It got away from me and ran over my leg! Finally I did get it going, but I was two laps behind

everyone. My golden opportunity to make a decisive leap in the point standings was lost in one stupid move, but perhaps a certain Calvinist notion of retribution had been satisfied.

In a rainy British Grand Prix at Aintree I was second, again to Trips, but I had a horrible fright. I hit a groove of deep water going into Melling Crossing and I could see my right front tire just stop turning. Hydroplaning was not as common then as it became with the advent of wider tires, but it was just as debilitating. I had no steering and I was headed for this meter-thick gate post with that image of the tire's stationary tread pattern etching itself on my retina. Finally my wheels cut down through the water enough to get a grip and I missed the post by a hair. That ruined my concentration. A few years earlier it would have been forgotten—like a letter dropped in a mailbox— the instant that the wheel caught hold, but by 1961 it stayed with me.

The Nürburgring followed and again I screwed up. Ferrari had still held off on designating a Numero Uno and Trips and I battled for second behind Moss. I was ahead on the long straight to the finish when we ran into this wall of water from a sudden rain storm the Eifel

Mountains are given to and both of us nearly crashed. Trips got straightened out first and scrambled across the line ahead of me.

Since the championship was determined by points from your best five races, there were all sorts of mathematical possibilities going into the Grand Prix of Italy at Monza, but certainly if either Trips or I won and the other didn't finish there would be no need for Ferrari to choose a Numero Uno—one of us would be World Champion.

I was later told that Ferrari watched the race on TV and that when Ginther, Rodriguez and Baghetti went out and Trips had crashed (there was no immediate report on the extent of his injury) Ferrari had said: *"Abbiamo perduto.* We have lost." But I won. And I won the championship, which was a warming relief, a soaring feeling. But it was to be a short flight. When the race ended I asked Chiti about Trips. He muttered something but I could tell from his face that it was not the truth. I suspected the worst, but it was not until after champagne and congratulations on the victory stand that I was told. Fourteen spectators had died as well.

After Trips' death I stayed around Modena for several days

231

Monza winner

At Monaco in 1961, Phil leads McLaren and Graham Hill.

On the way to victory in the 1961 Belgian GP.

because Tavoni asked me to. One day I was sitting over coffee in the little hotel opposite the factory in Maranello and Ferrari came across looking fatigued and hollow-eyed with the shadow of a beard. Since the accident, government officials and the press had put him through the mill trying to fix the blame for the disaster on the car. The Italians love to do that. The one thing that can quickly unite the Catholic establishment and the Communists in Italy is the immorality of racing. After a fatality, they seek to place the blame on the car and the venal men who build it. That was particularly unfair to Ferrari because never did he allow engineers to compromise safety for lightness. Ferraris, as I've said, were never fragile. We did not have steering columns break, hubs fail or wheels fall off. "With Ferrari you have not to worry . . ."

There was, though, something about the ambiance at Ferrari that did, indeed, seem to spur drivers to their deaths. Perhaps it was the intense sibling-rivalry atmosphere Ferrari fostered, his failure to rank the drivers and his fickleness with favorites. Luigi Musso died at Rheims while striving to protect his fair-haired-boy status against the encroaching popularity of the Englishers—Peter Collins and Mike Hawthorn. And Peter Collins, a firm favorite while he was living in the little hotel within earshot of the factory, began to get a Ferrari cold shoulder when he married Louise King and went to live aboard a boat in Monte Carlo. Peter was dead within the year. Time and again I felt myself bristling as Ferrari used Richie Ginther and Dan Gurney to needle me. And certainly Trips and I were locked in direct combat.

Not that Ferrari consciously teased his drivers into foolhardy excesses or urged them to perform beyond their abilities. We were all adults, all responsible for ourselves. We willingly jumped into the cauldron that he kept at a boil.

It pained me to see Ferrari raked over the coals for a reason most distant from reality, and after several days of his ordeal by press and government I went to see him at home in Modena and he asked: "Will you come back next year?" and I said, "Yes. Yes, I will." It was an emotional moment. Ferrari was capable of turning on great displays of emotion and one could not say that they were phony—any more than the third act of *Tosca* is phony. But there was something essentially theatrical about these displays—a great outpouring and then the curtain fell and it was over. La Scala might have lost a star when Ferrari went into cars.

My haste in telling Ferrari I would return in 1962 turned out to be ill-considered. I started having second thoughts and, believing they had to do with money, asked for an increase before I would sign. Money had never amounted to much at Ferrari. It took several years before I worked up to three or four hundred dollars a month plus half the prize money—a joke by today's standards. But still there was that matter of feeling "at home" at Ferrari, and I didn't want to risk the newness of BRM or anywhere else.

It was only after I signed the contract that the bombshell dropped. I learned that all the key people had left Ferrari and gone off to Bologna to form a new company, ATS. With them went Tavoni as well as Chiti. What was left were junior engineers and personnel. The team manager was to be a man named Dragoni, which was bad news because he disliked me intensely and I returned the sentiment. Nor did I get along well with the engineer, Forghieri, either.

Again I was back to Square One. Again no one would believe me when I told them what was wrong with the car. They attributed my every complaint to causes other than a faulty car. A pet theory of Dragoni's, which he never relinquished, was that I had been "impressioned" by Trips' death. Nor did he keep his notions to himself. Right off the bat at Aintree it was clear that the new car was a disastrous pig. And to compound my troubles I had a terrible case of flu, the effects of which were still with me weeks later at the Targa Florio in Sicily.

Dragoni started his game of pit psychiatrist and informed me that my "flu" was nothing else but a psychological reaction to Trips' accident, which I felt was not only incredibly presumptuous of him, but wrong. I knew where I was psychologically and I was no more nor less "impressioned" by Trips' death than I had been by Collins' or Portago's or Musso's or Behra's or Hawthorn's. The list goes on. Actually Mike's had perhaps affected me more because it had happened on the highway and not in a race.

At the Targa during practice I had the only serious accident of my racing career, though I was not hurt. The throttle stuck on my car and I literally flew off this small cliff. But Dragoni would not believe me. He insisted that it was my "hysteria" over Trips that had caused the accident. So it was with no small satisfaction that I heard, after the car had been put back together and Bandini was driving it at Nassau, that he drove right off the first turn—with a stuck throttle.

Nothing improved, neither my relationship with Dragoni nor the car. Still, I was tied for the championship as late as Zandvoort. My car was demonstrably slower on the straights, but they would not admit to the obvious. Forghieri had come up with the idea of a wider track (he had probably fought with Chiti about it the year before and been overruled) and he was trying out all his notions on my car. The handling was improved but wind resistance increased. It was so bad that the car was down several hundred revs on the straights of faster circuits. I had to slipstream guys like Rodriguez to keep up with traffic. But nothing would convince them. "What's wrong with you?" they'd ask.

After Zandvoort it was on to Spa, another high speed course where top speed spells the difference and where the wide-track car still couldn't get out of its own way. The race had ended and it was night and I happened to overhear Dragoni who had just got his call through to Modena. "*Commendatore? Si. Si . . . .Si. Si. Ma il tuo grande campione non ha fatto niente. Niente.*" ("Your great champion didn't do a thing.") I thought: "You son of a bitch." That tore it for me.

If the year was grinding down with frustration, it must have been miserable for Ferrari, too. The cars were clearly inferior to last years's cars at a time when he needed to prove to the men who had left how well he could get along without them. A scapegoat was needed and I

234

Hill leads the opening lap at the 1961 British GP.

was elected. Since his cars were so directly an extension of Ferrari's own being, to admit fault in them was to admit fault in himself. That was something he could not do. As I look back now, all my years at Ferrari seem to be a macabre dance, by everyone concerned, to avoid blame. The mechanics, the engineers—yes, even the drivers. No one wanted to be stuck with *mea culpa*. I think Ferrari's personality—the demanding, accusing, unforgiving father—had a lot to do with all that scrambling to armor one's ego.

My relationship with Ferrari ended more with a whimper than a bang—there was never any face-to-face final scene. After Monza, where they had screwed up my engine to the point even they recognized it, I agreed to meet Chiti in Monza and discuss my joining ATS. It was an emotional rebound and should have been more carefully considered. I was in a pattern of behavior that's like being behind a skid with your steering—the motions I made now were the motions I should have made then, and all I was doing was making matters worse.

At the time it did not seem like such a bad move. ATS had big plans and was well-financed. They had built a factory, an engine and a racing car in an incredibly short span of time. But the car was giving

The pits at Rheims, 1961. The Rodriguez brothers are on the wall.

problems with excessive fragility; then Italy's first major financial crisis since the postwar recovery struck, and some backers pulled out, and everything came down like a house of cards.

Ferrari took back the mechanics and the little guys who, I suppose, he figured hadn't known any better. Tavoni ended up with the Auto Club of Italy in Milano and Chiti went back to Alfa. I went to Cooper, which was another disastrous move on the string I had going. Anyway, I wrote "The End" to my relationship with Ferrari, or—more accurately—"*La Commedia e Finita.*"

Time has lent perspective to my feelings about Ferrari. I realize that then I was whirling so tightly in the orbit of my own problems that I was unaware of the real greatness of the man. Our problem was one of communication that had nothing to do with language differences. I responded to him from the bias of my own needs and he, in the jargon of today, could really push my buttons. Even now, years later and miles away, he lurks in my unconscious and surfaces from time to time, once in an amusing, amazing dream I had in 1971.

The setting has a Latino feeling—like Mexico or South America. I am in a big marketplace and I have the feeling there are race cars nearby if only I could find them. There are all these people I know involved in racing. Suddenly there is Dr. Camissa. (He had been associated with the Mexican road race as sort of a liason between the local Italian community and the Italian racing people.) We exchange pleasantries and then I see Ferrari. "Ferrari is here. What is he doing here?" I ask. Dr. Camissa says, "I don't know. This is really something. He never comes to races." As Ferrari approaches us, I'm thinking: "I wonder if he's seen the article in *Motor Trend*?" (Ed. note: a frank interview with Hill published in September, 1970.) Ferrari comes closer and closer: *Filipo, caro mio! Ti abbraccio. I embrace you,"* And I sigh with relief. "He liked the article," I say to myself.

Then suddenly Ferrari pushes me away and begins the most abusive tongue lashing imaginable. "How could you? Do you really think I am such an awful person?" etc. I am trying to tell him: "Look. I said what I felt in my heart." But he is getting madder and madder and finally in his anger falls over backwards and cracks his head on the pavement. Dr. Camissa gets all upset, and I'm moaning "Oh, my God. Ferrari. Is he alive?" We run to him and while he's lying there he starts shrinking up, smaller and smaller, until he's only two feet long.

I say "We've got to do something!" Camissa is saying "Call an ambulance!" I lift Ferrari up and start to carry him and suddenly he comes to—the impression I get is that he has been faking all along—and he bites me on the arm so hard I have to drop him!

Then I woke up. Make of that what you will.

Rheims, 1961.

Von Trips at Rheims, 1961.

ENERGOL **BP**

Monaco GP,

Hill and Mairesse at L'eau Rouge hill during the 1962 Belgian GP.

Now that it is all distant in time and place I have a different feeling for Ferrari. I do not try to separate what I consider the Good and the Bad in him. I accept him whole, recognizing that, as in many people of genius, his great virtues are the other face of his faults. I now have a more rounded appreciation of him. And, too, we have both mellowed. I had a nice meeting with him in 1974 when I saw him at his new track in Maranello. I was embarrassed at first because there had been such a long time when we would not even speak because of the unpleasantness with which we had parted, but we had a pleasant talk. It was the first time that I personally had ever experienced him as gracious and warm.

Ferrari is still actively involved in racing despite his seventy-six years. There are few people in any endeavor who continue to work on the same energy level for so long as Ferrari has done. Some symphony orchestra conductors are the only ones who come to mind. In the automotive field maybe Bugatti and Henry Ford could match him. Think of it. Here was a man who grew up within miles of his birthplace, who never left the provinces for the seats of power in the city and yet earned a world-wide reputation that brought the great trekking to his door. (And waiting in his office.)

Ferrari is the last of a type, the individual industrialist, a man with a personality so strong anything he touches is marked by it. He wanted to do something memorable on his own, something identifiably his and his alone. He did. I cannot imagine that in this conglomerate-ridden world anyone like him will ever exist again. His era is ended and he has left his mark on it. ⌖

239

Formula One Ferraris

Illustrations by Ken Rush

When he was fourteen in 1945, Ken Rush–who remembers being able to draw before he could walk–became the youngest artist to have a painting accepted for the Royal Academy Summer Exhibition. Since that painting–of rooftops seen from his bedroom window in Luton, England–Ken Rush has pursued a successful career in commercial art and has become a well-known specialist in the demanding field of mechanical and automotive illustration. His work has appeared in numerous publications and his recent book (with text by his wife Jean)–The Fantastic Book of Machines–has been translated into seven languages. On the strength of that book, Mr. Rush was chosen to represent Great Britain's illustrators in European Illustration, 1974-75. In this chapter, he depicts the development of Ferrari's Formula One cars, basing his pictures on contemporary photographs.

The cars shown here–graceful, purposeful, functional, eventually space-age in their complexity–can be viewed as a portrait of some of the best in postwar race car technology. And it seems only fitting that, after over a quarter century's participation in motor racing's ultimate arena, Ferrari is once again a serious threat in Formula One. As this book goes to press, Niki Lauda has won a smashing four Grand Prix races and stands first in line for the 1975 World Championship. Whether he wins it or not, I know of no one who would deny that Ferrari's resurgence has been good for the sport. As for Ferrari's great contribution to GP history, that is perhaps best understood by simply looking at the cars. (Ed.)

1948 Tipo 125

In Ferrari's Formula One debut—the 1948 Italian GP at Turin—Raymond Sommer drove this car to third place. Ferrari's first GP racer was powered by a 1500 cc engine equipped with a single-stage supercharger and had torsion bar rear suspension.

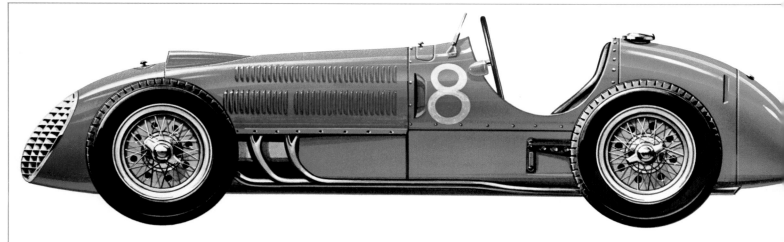

Alberto Ascari won the 1949 European GP—held at Monza—in this machine. Improved over the earlier 125's, it had two-stage supercharging, two cams per cylinder bank and swing-axle rear suspension to improve handling.

1949 Tipo 125

1952/53 Tipo 500 F2

Alberto Ascari won the Formula Two World Championship twice in a row driving a 1980 cc four-cylinder car similar to this.

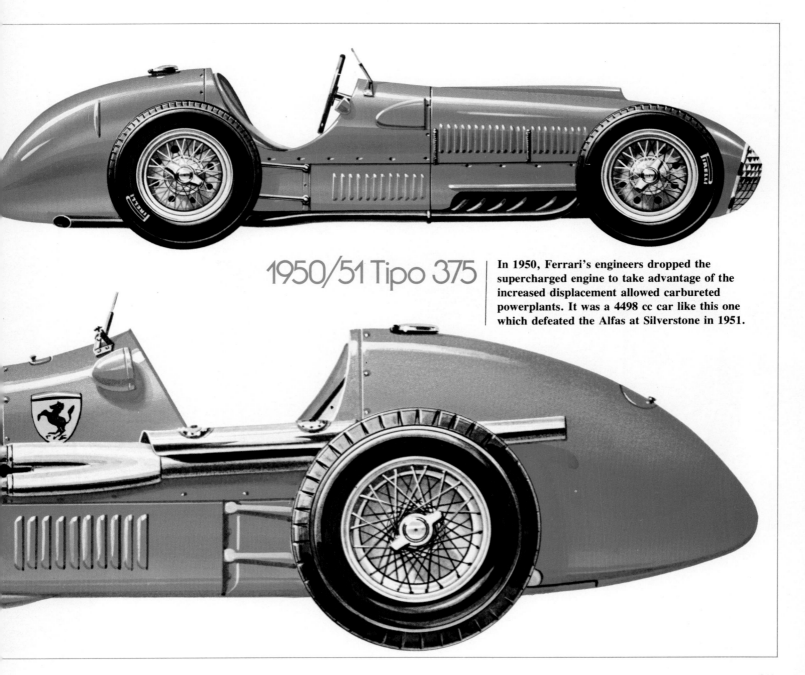

1950/51 Tipo 375

In 1950, Ferrari's engineers dropped the supercharged engine to take advantage of the increased displacement allowed carbureted powerplants. It was a 4498 cc car like this one which defeated the Alfas at Silverstone in 1951.

1954 Tipo 553 Squalo

Two of Mike Hawthorn's cars are pictured here, both with four-cylinder engines derived from the 500 F2 but now displacing 2.5-liters. In his 625 (left), Hawthorn maintained second place for half the Rouen GP but was then forced out with a broken oil line. He had better luck driving the Squalo (above) and won the Spanish GP in this car though it was otherwise an unsuccessful, if beautiful, design.

1954 Tipo 625

An unhappy development of the 1954 cars, Ferrari's Super Squalo was not really competitive, unpopular with the team drivers and never won a Grand Prix race.

1955 Tipo 555 Super Squalo

1956 Lancia/Ferrari

When Lancia handed over its Formula One cars to Ferrari at the end of the 1955 season, Ferrari once again had a competitive racer. In the car pictured here, Peter Collins won the Belgian GP.

1957 Tipo 801

Ferrari's much-modified version of the Lancia-designed GP cars had a redesigned frame of larger diameter tubing, revised suspension and detail changes to the V-8 engine which—as in the previous season's car—served as a stressed member. Peter Collins drove this car at the 1957 Monaco GP but was one of three Ferrari drivers eliminated by accidents.

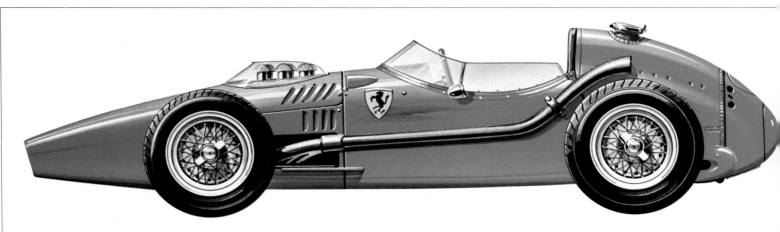

Forceful, powerful, the 2.4-liter, V-6 Dino was a great success and Hawthorn won the 1958 World Championship in one of these cars.

1958 Tipo 246 Dino

1959 Tipo 246/256 Dino

Not as successful as its predecessor, the Dino for 1959 was still a smooth but aggressive-looking car which could not be counted out. In the European GP at Rheims, Tony Brooks drove this car to victory.

The last front-engine Ferrari GP car, this Dino was tough, reliable and thoroughly outclassed by its more modern competition. It represented the end of the line for Ferraris of its basic type and after its only win of the 1960 season—at Monza with no rear-engine British cars in the race—Ferrari passed up the U.S. GP to concentrate on preparing the new 1961 racing car.

The dominant, most powerful car of the new 1.5-liter formula during 1961 was Ferrari's Tipo 156, a car which looked right and was right. Phil Hill drove one to that year's World Championship and Wolfgang von Trips finished second to Hill at the Belgian GP in this car.

1961/62 Tipo 156

1960 Tipo 256 Dino

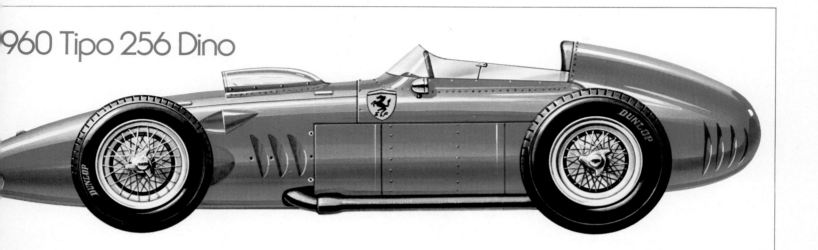

1963 Tipo 156B

Lighter and more compact than its immediate forebears, the dull-looking 156B was quick but not especially reliable. Perhaps its most interesting features were its monocoque rather than tube chassis, Bosch fuel injection and its cast alloy wheels, major changes for Ferrari.

With an improved chassis, new V-8 engine and graceful bodywork, the 158 again made Ferrari a serious threat in Grand Prix racing and John Surtees won the World Championship. He drove this car—painted NART colors—to second place at the U.S. GP.

1964 Tipo 158

1965 Tipo 1512

Powered by a compact 1.5-liter engine of brilliant design and execution, the 1512 was a relatively successful car in its first season but its development was cut short by the introduction of the new 3-liter formula. Lorenzo Bandini finished second at Monaco in a 1512 and drove the car shown here to eighth place at the French GP.

1966 Tipo 312

The good-looking 312—first of a long-lived model series—was powered by a strong V-12 engine with white-painted exhaust pipes, something new for Ferrari. Although team politics made life difficult for John Surtees in 1965, he won the Syracuse GP and the Belgian GP in the new 312. The car pictured here was Ludovico Scarfiotti's Italian GP winner.

Engine modifications including three-valve heads and two camshafts per cylinder bank together with different bodywork characterized the 1967 312. Lorenzo Bandini was in second place at Monaco in the car shown here when he was fatally injured in an accident on the 82nd lap.

1967 Tipo 312

1968 Tipo 312

Lighter than the 1967 car and with a more powerful four-valve per cylinder engine in an attempt to counter the Cosworth V-8 threat, this 312 also mounted an hydraulically operated rear spoiler. Jacky Ickx won the French GP in the car shown here, its best finish for the year as it was not overly fast or reliable.

1969 Tipo 312

Except for a third place at Zandvoort, this year's
312 proved uncompetitive and unreliable. The V-12
had reached the end of its development cycle. Chris
Amon drove this car at the French GP but retired
with a blown engine.

1970 Tipo 312B

The introduction of a new flat-twelve which, by mid-season, developed some 460 hp at 11,600 rpm put Ferrari back into contention. The 312 B (for Boxer or horizontally opposed) was a sturdy car with bodywork designed in a wind tunnel. Jacky Ickx won the Canadian GP in the car illustrated here.

With revised, even more efficient bodywork, a roller-bearing crank which permitted 12,800 rpm and other detail changes, the 312 B2 should have been more successful than it was. Although Mario Andretti won the South African GP in this car, the remainder of the season was vastly disappointing for Ferrari thanks to frequent engine breakdowns.

1971 Tipo 312 B2

1972 Tipo 312 B2

With different aerodynamics, suspension changes, and minor detail improvements, the 1972 312 B2 performed, if not brilliantly, at least better than the previous year's car. Jacky Ickx scored a fine win in the German GP. The car seen here finished fourth at the Argentine GP with Clay Regazzoni at the wheel.

1973 Tipo 312 B3

The first of a new generation car conforming to new safety regulations, the 312 B3 was a full monocoque with no tubular stiffeners and the engine acting as a stressed member. All the car's components were attached to one or another of its three major bulkheads. Air ducts to oil and water coolers were revised during the season resulting in several appearance changes. This enormously complex racing car performed indifferently, the version seen here finishing seventh in the Austrian GP where it was driven by Arturo Merzario.

1974 Tipo 312 B3

More new bodywork, detail changes and refinements enabled the 312 B3 to begin living up to its promise. The car was now truly competitive and the one shown here was driven to victory in the Spanish GP by Niki Lauda.

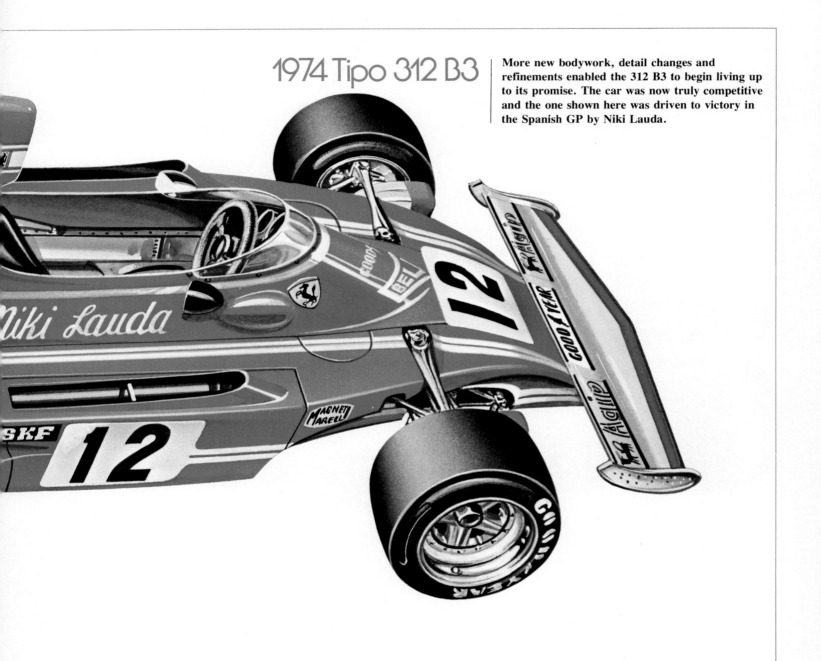

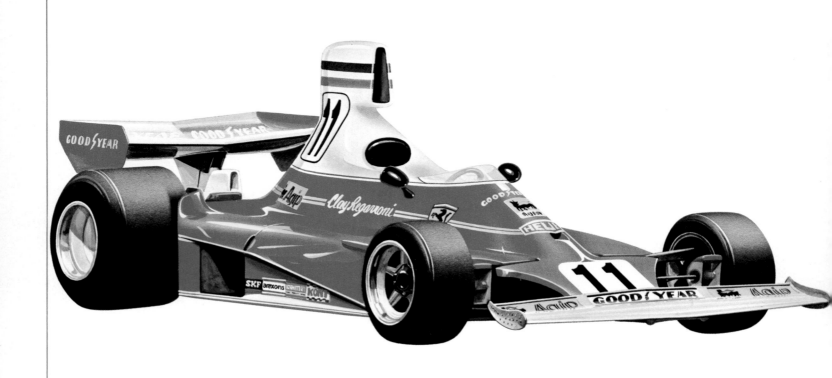

1975 Tipo 312T

New bodywork and front suspension and a transverse five-speed gearbox are highlights of the latest Ferrari Formula One car, the 312T. Since the South African GP where Regazzoni was halted by a broken throttle linkage, and Lauda placed fifth, the car has proved overwhelmingly competitive, a winner.

CHAPTER 12

"Il Grande John"

by Denis Jenkinson

Few automotive writers understand as much about Grand Prix racing drivers and their unique skills as Denis Jenkinson. Onetime Grand Prix motorcyclist and later sidecar passenger for Belgian rider Marcel Masuy and World Champion Eric Oliver, he eventually quit two- and three-wheel competition to become an automotive journalist. He was four times a passenger in the Mille Miglia and rode with Stirling Moss in a Mercedes Benz 300 SLR when Moss won the race in 1955. Out of these firsthand experiences came Mr. Jenkinson's classic book, The Racing Driver, *and his detailed study of GP racing during the Fifties,* A Story of Formula One. *In this chapter, he traces the racing career of John Surtees, a man who was World Champion on motorcycles before he turned to cars and won Ferrari's last Formula One World Championship.*

S ince the 1930's when Enzo Ferrari ran a motorcycle racing team—equipped with Rudge and Norton machines—as well as his Scuderia Alfa Romeos, he has been appreciative of the merits of top-flight racing motorcyclists as potential racing drivers. Naturally he noted the influx of British riders to the Italian teams of Moto-Guzzi, Gilera and MV-Augusta during the Fifties and he singled out Geoff Duke as a natural star who would do well in racing cars. But Duke was not very interested in the idea and though other works riders like Fergus Anderson, Ken Kavannagh and Keith Campbell made forays into the automobile racing world, Enzo Ferrari did not see what he was looking for among them.

At Monza in 1964,
World Champion Surtees and Ferrari.

**Surtees is shown here at Monaco in 1963.
He finished fourth.**

Then, in the late 1950's, Ferrari took note of a serious-minded young Englishman named John Surtees who was riding for MV-Augusta and winning just about everything in sight. Surtees—who was to capture seven World Championships as a motorcyclist before he quit two wheels in 1960—had made it clear by 1958 that he wanted a chance to drive Formula One.

"For several years I'd been doing the same thing over and over again," Surtees once told me of his motorcyling days. "I was anxious for fresh fields and was actually considering going into business before I met Tony Vandervell, Reg Parnell and Mike Hawthorn. That's how I first got into motor racing."

Like Ferrari, Vandervell was quick to see the potential in Surtees as a driver and although Vandervell had withdrawn his team of beautiful Vanwalls from racing after the 1958 season—on doctor's orders as his health was suffering—he continued to dabble at his favorite pastime. He gave Surtees the opportunity to extensively test one of the Vanwalls and get some experience behind the wheel of a Grand Prix car, hoping eventually to return his team to competition.

As things turned out, the experience Surtees gained with the front-engined Vanwalls—and the Aston Martin Formula One car which he also tested—was of little avail. Tony Vandervell never did restart his team and the Aston Martin was far from ready. But in early 1960, Surtees joined Team Lotus for the Monaco GP and made a good impression with the fragile rear-engined 2.5-liter Coventry-Climax powered car before it failed. However, John's strong personality clashed with that of Colin Chapman and Surtees left Lotus—having done well enough to finish ninth in Championship points by season's end—to join Reg Parnell's Yeoman Credit team of Coopers in 1961.

When Surtees began his driving career, casual observers who did not know just how serious and methodical he was, were apt to dismiss him when they heard of a spin while testing the Vanwall. Some put him in the "fast but wild" category with little future. But John knew exactly what he was doing—he had deliberately driven the Vanwall over the limit to learn what it was like to spin a GP car—and explained his tactics to me this way: "I was trying to gain four years of experience of handling a Grand Prix car in a few weeks when I began, so that I could hopefully start on something of an even footing with my opponents."

All the while, Enzo Ferrari was watching John's progress from his lofty Modena vantage point and what he saw interested him. He realized that his ideas about racing drivers were similar to Tony Vandervell's, and that the English industrialist believed in Surtees' potential. Also John's own personal philosophy would have meshed perfectly with Ferrari's. "Whatever you do needs a competitive spirit,'' John believed. ''When there's a challenge, there's no time for sitting back.'' Besides, Surtees' standing in Italian sporting circles had always been high—after his MV-Augusta years—and Ferrari knew that the inclusion of this Englishman in his Grand Prix team would be welcomed by his compatriots, however nationalistic they might be. He approached Surtees with an offer of a place on the Scuderia Ferrari Grand Prix team for 1962. Presented with this great plum, Surtees . . .refused.

John's decision was characteristic of his honest approach to things—he hates artificiality in any form—but though he felt honored by Ferrari's offer, he wanted to do one more season before joining the Italian team. Whatever his native ability as a driver, John wanted to feel more ''complete'' too and that meant more experience. He spent the 1962 season with the newly-reorganized Yeoman Credit-Bowmaker

Team, helping to develop their new Lola, a trying if rewarding experience.

The year he spent with the Lola was to prove a significant factor in John's ultimate future, though he probably did not realize it at the time. During that year, he became heavily involved in the Lola's development. Even though he did not have an engineering degree, John was already deeply involved in the design and construction of racing cars before he joined Ferrari. Perhaps he was more involved than was good for his driving. Still, his fascination with racing's mechanical side was understandable for he had, as a youth, spent a five year apprenticeship with the Vincent-HRD company where he had learned the basics of mechanics and engineering.

But although John had ample opportunity to indulge his interest in design and development at Lola, he entered a wholly different set-up

onaco, 1963.

at Ferrari when he joined the team in 1963. There he found an army of designers, engineers and technicians in the racing department, all of whom accepted that his role would be that of driver only. Enzo Ferrari had always believed that his technicians could produce a car that would win, and that all his drivers had to do was to drive it. And indeed most of the drivers he had employed over the years were happy in that role. But not John Surtees. He continued to take a deep interest in racing car technology. Though he was not allowed to interfere with the design department, some of the Ferrari engineers welcomed his analytical approach to driving and felt that it helped them in their development program.

Surtees return to Italy was greeted with great joy by the *tifosi*, for his popularity rating with the Italians when he was riding for MV-Augusta had been very high. He had immersed himself in the Italian way of life completely, and learned to speak Italian, eat Italian and think Italian, all of which endeared him to the sporting populace. It was not for nothing that the Italian press referred to him as *Il Grande* John—John the Great (not Big John, as the English press misinterpreted it). They did not do this simply because he won World Championships for MV-Augusta, but because they looked upon him as "one of them," unlike so many riders and drivers who have raced for Italian factories and kept themselves aloof from the Italian way of life.

When he joined Scuderia Ferrari, Surtees knew the sort of people he would be working with and accepted them the way they were and fitted into their scene. He had always been happy living in Italy, so it was no hardship to make Modena his home while he was racing. And the very fact that he was always on hand was greatly appreciated by the Ferrari team, his relationship with them being amicable from the start—all of them, that is, except one: the new team manager.

Eugenio Dragoni, an old acquaintance of Enzo Ferrari, had taken on the job when Tavoni quit. Dragoni was from Milan, an elderly man involved in the pharmacy and cosmetics business, who had time on his hands and wanted to be part of the organizational side of Italian motor racing as a hobby. He was fiercely patriotic and inspired by the feudal Italian mentality that believed each Italian town was really an entity unto itself.

Consequently, Dragoni really had no great admiration for Modena, an attitude that was to have far-reaching consequences as another rising star on the Ferrari team was Lorenzo Bandini. And Bandini had been a garage mechanic in Milan since age fifteen. Dragoni soon made it clear that he considered him his protégé, and could see no reason why Bandini should not become World Champion at the wheel of one of Ferrari's excellent Grand Prix cars. Thus, the inclusion of John Surtees on the team was received with little enthusiasm by the team manager. Bandini, it must be said, remained unaffected by all this and had a high regard for *Il Grande* John.

In complete contrast to his team manager, Ferrari himself was one

Dutch GP, 1963. Surtees placed third.

of Surtees' greatest supporters and made great play of the fact that they got along extremely well together. Normally Ferrari would sign up drivers and treat them as paid chauffeurs, expecting nothing more than that they should always be at his beck and call and should win races with his cars, whether they be Grand Prix cars or sports cars.

With Surtees it was very different, and Enzo allowed himself to be interviewed by the press on the subject of *Il Grande* John, and also permitted numerous photographs to be taken of the two of them in happy accord, chatting by a car, looking at things in the racing shop, even lunching together in fashionable Modenese restaurants. The press found this a new experience, for Ferrari was usually very distant and elusive, preferring to maintain an air of mystery rather than open pleasantness. The whole of the Italian sporting world was viewing the arrival of John Surtees in the Ferrari team as ''the return of the prodigal son,'' and Enzo himself was encouraging this feeling.

In 1961, the Ferrari team had dominated Grand Prix racing; but the following year, when the British teams got their teeth into the new 1.5-liter Formula, things were different. Nineteen sixty-three saw Scuderia Ferrari preparing with renewed vigor and enthusiasm. But sports car racing was then very important to Ferrari's sales program, so a lot of effort was put into the long-distance racing campaign, while the Grand Prix effort was generally concentrated on a lone entry for Surtees. This was not as strange as it might seem, for among the

Ferrari drivers John was really the only one whose driving ability made him capable of beating the opposition. He teammates were Lorenzo Bandini, Willy Mairesse, Ludovico Scarfiotti and Michael Parkes, and the fact that Bandini did not figure in the Ferrari Grand Prix team during the first part of the 1963 season did nothing to lessen Dragoni's antagonism towards what he called the ''favored Englishman.''

In the long-distance sports-car races Surtees was partnered variously, sometimes by Bandini, sometimes by Mairesse, sometimes by Parkes, but from the outset it was clear that he was head-and-shoulders above the others in driving ability. No matter who partnered Surtees in the sports car races, they were unable to match his pace, which made his relationship with teammates a difficult one. He was extremely tolerant of the situation, when other drivers might have thrown a fit of temperament, and though he grumbled as he watched his lead being lost while his teammate was at the wheel, he soon made amends when his next stint came.

Numerous victories fell to the Ferrari sports car team with Surtees and one or other of the team drivers at the wheel. Naturally he would have liked a strong ''supporting cast'' but the team was settled for the season so he made the best of it. In Grand Prix racing he was on his own most of the time, battling valiantly against numerically superior odds, but always giving of his best, and always in the hunt, though victories were sparse.

the 1963 Belgian GP, fuel injection woes ended Surtees' drive.

In 1964 Ferrari produced a new Formula One V-8 engine which
put John on an equal footing with any of the opposition, and he began
to use this benefit very quickly. As a Grand Prix driver he was
emerging in the top echelon, the disappearance of Stirling Moss in
1962 leaving a void at the head of the Grand Prix scene that was
contested by four drivers, all striving to replace the accepted
"master." These were John Surtees, Jim Clark, Graham Hill and Dan
Gurney. For a long time there was little to choose between them, but
gradually Jim Clark developed a marked superiority which put him
ahead and into a class of his own. John Surtees also improved, but not
to the same extent, while the other two appeared to have reached their
zenith.

With the strength of the Ferrari team behind him, John drove
brilliantly throughout 1964, gaining the World Championship for
himself and Ferrari. There had been a degree of good luck which
influenced the result, but equally there had been a certain amount of
bad luck, which made the task more difficult than it should have been.
However, it was the end result which mattered, and the history books
record for all time that John Surtees was the 1964 World Champion of
Grand Prix driving, in a Ferrari car. It had been a good year for
Ferrari generally, as his V-12 sports cars had their usual sweeping
success in long-distance racing. For Surtees, his win in the German
Grand Prix on the long and arduous Nürburgring was an outstanding
race, while his victory in the Italian Grand Prix at Monza before a
very large partisan crowd ended in complete chaos as the *tifosi*
acclaimed *Il Grande* John. When Lorenzo Bandini finished in third
place, the Milanese supporters went wild, and everyone in the Ferrari
team was deliriously happy. Even the usually sour-faced Eugenio
Dragoni was reasonably pleased, but he made it obvious that he would
rather have seen the positions of the Ferrari drivers reversed.

By 1964, Ferrari engineers had already produced a new
horizontally-opposed twelve-cylinder Grand Prix engine, but until it
was proved completely race-worthy, Surtees preferred to use the V-8,
leaving the "prototype" driving to his young teammate. This was
reasonable enough, for John was the pacemaker and natural winner of
the team, so it would have been a waste of talent to make him drive
the newly-powered car before it was really ready. But signs of friction
between Milan and Modena could already be felt as Dragoni gave
interviews to his local press friends, decrying the worth of Surtees and
praising the prowess of Bandini, pointing out that he was under the
handicap of having to play 'second-fiddle' to the Englishman. "Any
right thinking Italian," he said, "should be able to see that in Bandini
Italy has a true World Champion, if only Ferrari would play fair with
him." To his credit, Bandini did not endorse these views, knowing
full well that he could not out-drive Surtees, let alone some of the

ari and "Il Grande John"—a luncheon party with the prodigal son.

In the racing department at Maranello.

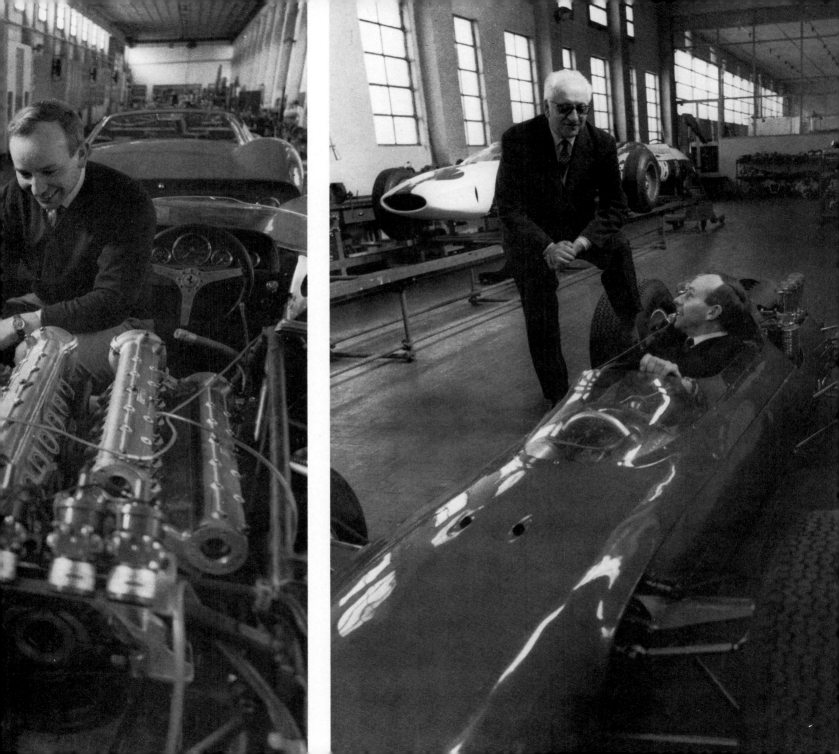

Flying low at the Dutch GP, 1964. He finished 2nd.

opposition, but he was confident in the knowledge that he was the best Italian racing driver.

In the sports car events Dragoni continually paired Bandini with Surtees, in the hope that his young protégé would have the benefit of the faster driver in the results. Loyal as he was to the Ferrari team, John Surtees could not conceal his displeasure at having the young Italian as his co-driver, continually watching the lead being lost while Bandini tried to maintain the pace. It was obvious that there was a big gap between the two drivers' abilities, and at times poor John got quite exasperated by the situation. Though he accepted it as part of team orders he was not happy and he could, of course, find no level of accord with Dragoni.

To Enzo Ferrari, who was merely viewing the results from his distant position in Modena, never actually attending a motor race, the situation seemed highly satisfactory and his team manager seemed to be getting the desired results and doing a good job. Internal friction within the team was of no importance; in fact, he often deliberately caused it if he felt there was an air of complacency among his drivers. It was his philosophy that a certain amount of ''needling'' of the drivers would keep them on their toes and make them try harder.

In 1965 the team remained ostensibly the same, but John got arrangements changed in the sports car races, taking Ludovico Scarfiotti as his partner in place of Bandini. This did little to improve the situation though, for Scarfiotti was not as fast as Bandini. The thirty-two-year-old Scarfiotti was an amiable fellow, however, and did his best but there was little hope of winning any of the long-distance races.

They were really not at all a happy bunch of drivers in the Ferrari team by 1965. Mike Parkes had little time for John Surtees and rather resented his presence in any form other than a driver. Parkes was a qualified and talented engineer, employed by Ferrari in that capacity, but allowed to drive in the sports car races as an ''extracurriculum'' exercise. He became exasperated when John—drawing on his Lola experience—tried to involve himself in the racing department's technical side. In addition, the two men's upbringing, education, and their ways of life, were poles apart. Though they were two Englishmen working for Ferrari and living in Modena, Surtees and Parkes might just as well have been complete foreigners to one another.

It was not surprising that at times John felt he was forging a lonely path with the prancing horse team, and though Ferrari was obviously behind him it did not seem to be having much effect. In all conscience he was doing his best, always driving and working for Enzo Ferrari and not for himself, as a good team-man should do. If those around

At Brands Hatch in 1964, Surtees placed 3rd.

him seemed to be letting personal feelings affect their efforts for the team, he despised them for it. John Surtees was always very outspoken on the subject of team spirit and team discipline, maintaining that his approach to everything was the right one. He could never be charged with not doing his best for the team, but a critic later on said of him, "John's idea of the perfect team is one in which Surtees is the owner, Surtees is the designer, Surtees is the engineer, Surtees is the team manager and Surtees is the driver; that way he could be certain of one hundred percent team effort from his staff!"

Ferrari's Grand Prix efforts in 1965 were not crowned with glory, for the friction and unrest was becoming more and more troublesome, spreading from the pits at the races back to the factory, and Surtees became conscious of thoughts and suggestions being put into Enzo Ferrari's mind before Ferrari had heard the full facts of the matter. There were times when Surtees was forced to make unnecessary journeys back to Modena to see Ferrari, merely to counteract the false impressions that he knew Dragoni was conveying back to the patron.

The new flat-twelve engine was beginning to show promise, but still was not fully race-worthy in John's estimation, so he continued to drive the V-8 engined cars, leaving the twelve-cylinder to Bandini. This was interpreted by Dragoni as a lack of interest in Ferrari's new engine, which was quite untrue. Honest, simple John merely wanted to race the Ferrari which he considered had the best chance of winning, for race winning was the sole objective. It did not help matters when Bandini's twelve-cylinder proved fairly reliable and finished races, while John had the earlier car break down on him. But the fact that he was right up with the opposition at all times, whereas Bandini was trailing along in mid-field, was not taken into account when Enzo Ferrari was being given the story of events by Dragoni. Later in the season, when he felt the twelve-cylinder was competitive, Surtees at last drove it. The car suffered continual breakdowns.

All his racing life John Surtees has erred on the cautious side as far as the mechanical aspect of things are concerned. He drove hard and earned the nickname "Fearless John." Through it all he never threw caution to the winds. Fast and safe has always been his dictum. His cautiousness over things mechanical can be blamed on his incomplete mechanical education, knowing only a certain amount and therefore being a little unsure of himself on engineering grounds. Had he not tried to become involved with the technical side of racing, and instead concentrated solely on driving, he would undoubtedly have been a far better driver than he was.

Surtees would have liked to have been taken into the technical side of Ferrari as an engineer, rather than a driver, but—because of the rigid Italian caste system of who held what job—there was no hope of

Monza, 1964—a tremendous victory.

279

this happening. Consequently, as the friction and troubles progressed during 1965 he began again to work with his friend Eric Broadley who ran the Lola Car Company. One of their projects was a Chevrolet-powered sports car for the North American series of races, to become known as Can-Am. An awkward situation was obviously developing. Whereas Ferrari had previously supplied cars for as much racing as John wanted to do, he now had to accept that his best driver was going to take leave to drive another make of car. As the Grand Prix season was drawing to a close, Surtees flew to Canada for a race at Mosport with the Lola-Chevrolet and there he suffered a monumental accident when part of the Lola T70's suspension broke. He was critically injured and on the verge of death, but he received little sympathy from either Ferrari or his Italian colleagues. They all believed he should not have been driving anything but Ferrari cars. Needless to say, Dragoni used the opportunity to stir up trouble among

his Milanese journalistic friends, and even while John's life was in the balance his racing future looked gloomy.

In addition to having a very strong will, John was also physically very tough, and he made a remarkable recovery during the following winter, being fit and ready to race again when the 1966 season began. Enzo Ferrari had kept John's place in the team open for him. Even before he was fully recovered—and this is typical of John's thoroughness—the first thing he did on returning to Modena was to drive a full-length Grand Prix distance on the Modena Autodrome test-circuit, using one of the 1965 cars. The Grand Prix Formula had started anew in 1966 with three-liter engines replacing the previous 1.5-liter ones, but Surtees was undeterred by the thought of having twice the power. In fact he was looking forward to it.

The Italian press was not at all optimistic about his future, however, goaded on by quotes from the Ferrari team manager, and

they doubted whether John was really fully recovered from his injuries. He knew well enough that he was fit, but any statements he made were put down to well-meaning bravado. Instead of getting a warm welcome back into the team, he returned amidst a feeling of doubt and suspicion. The fact that he had crashed in a car that had nothing to do with Ferrari caused feelings of doubt about his loyalty. Even his keenest supporters were prompted to wonder whether his true loyalties lay with Ferrari or elsewhere. The suspicions were soon swept aside, however, when the first race on the calendar took place. This was the 1000 kilometer sports car event at Monza and John demonstrated ably that he was right back on form and more than worthy of his place on the team.

With an effortless display of driving, he proved that he was by far the fastest driver on the track. In this race he was partnered by fellow Englishman, Mike Parkes. They were—strangely enough—well matched as racing drivers, even though Parkes' lap times were not as fast as John's. As a long-distance driver though, Parkes was very good, being consistent, reliable and gentle on the machinery, so that John had no qualms about his co-driver. In spite of atrocious weather conditions, and the windscreen wipers of their 330 P3 packing up early on in the race, they had an unchallenged run to victory, sharing the driving equally in stints of twenty-five laps at a time. Surtees did not have to drive to the absolute limit in order to maintain a solid lead, and in consequence Parkes was able to match his pace when he took over. It was a model long-distance race victory, all the more satisfying for John Surtees and his supporters, for the 500 kilometers that he drove in appalling conditions, proved that he was completely fit again and had lost none of his skill.

He followed this with an easy, virtually unchallenged victory in the Syracuse Grand Prix with the new V-12 Formula One car, the 312.

The Sicilian populace was overjoyed by the performance and Enzo Ferrari was well satisfied that *Il Grande* John was really back in the Maranello fold. While everyone else was praising John for his great comeback, though, Dragoni was making snide remarks that it was all very impressive, but anyone could see that Surtees was not really fit. And while he may have won the two early-season races, they were not against serious opposition.

Once again Dragoni was pushing Bandini forward for stardom, saying that Ferrari could not rely on Surtees anymore, for even if he was fit, there was always the possibility that he would go and drive another make of car, and "everyone knew what happened the last time he did that." Without actually saying as much, he was suggesting that Lorenzo Bandini was Ferrari's only real hope for the future; he was Italian, proud of the fact, and trustworthy. Dragoni did not add that most significant of all was the fact that Bandini came from Milan, but he might just as well have, for those who could see through the machinations of the old man knew him for what he was. There was little that Enzo Ferrari could do about all this, or indeed anything that he particularly wanted to do, for Dragoni was getting results in the form of race victories, and, ultimately, that was all that mattered.

At no time in his life has John Surtees looked the textbook picture of fitness, having a naturally pasty complexion and a worried expression. He seldom took time off to lie in the sun by a swimming pool and get a tan, nor did he strip to the waist the moment the sun shone at a race track, like drivers such as Stirling Moss, Jim Clark or Mike Hailwood did. He was physically fit and knew it, and saw no point in trying to convince anyone with a bronzed torso. However, Dragoni traded on this natural paleness of John's complexion.

When the team arrived at Monte Carlo, for the Monaco Grand Prix—over one-hundred laps of the street circuit—it was only natural that journalists who did not know John, or felt they could not communicate with him, asked Dragoni about the Englishman's fitness. Dragoni would shrug and say "He doesn't look very fit; I hope he can last one-hundred laps of this difficult circuit." And indeed, compared with some of the fit-looking and bronzed bodies around the pit area, poor John really did look as if he needed a Mediterranean cruise rather than a Mediterranean motor race. Those on the inside of racing, who knew him and had seen him drive at Monza and Syracuse, knew that there was nothing wrong with him; but the great majority of journalists at Monte Carlo knew little of motor racing and believed anything they were told, or anything that was suggested to them.

In addition to his negative attitude, Dragoni was being deliberately obstructive to John as regards the race itself. The new 312 was an unknown quantity on the twisty little Monte Carlo circuit, so Ferrari had sent along a "special" in the shape of the Tasman car that had proved successful during the winter. This was basically a 1965 car

Surt

with a 2.4-liter V-6 engine, a compact and very manageable car that John felt was just what was needed for the Monaco race. Dragoni thought differently, intimating that the new three-liter V-12 was the car which should be raced, the V-6 being merely a stop-gap.

A pretty open dispute developed between Surtees and Dragoni over this question of the V-6 versus the V-12. John's only criterion was that they had come to Monte Carlo to win, so it was up to him to use the machinery he thought had the best chance. Dragoni suggested that John's attitude was not helpful to the team, showing a lack of confidence in the new car. He refused to accept John's reasoning, hinting that the Ferrari designers and engineers would not be very happy when they learnt that the number one driver did not want to drive the new car.

At any other circuit this situation would not have arisen, for

at Monaco in 1965.

anywhere else the V-12 would have been superior; it was just that the freak conditions of the Monaco circuit looked as if they might favor the smaller V-6 car, and John's only interest was to win the race for Ferrari. After some heated words had passed between them, John was committed to race the new V-12 while Bandini took the V-6. The friction in the team that had been an undercurrent was now very much on the surface.

If the truth be really known, John was absolutely right in his estimation of the best car for the conditions, and Dragoni knew this. Insisting that John drive the 312 automatically gave Bandini a better chance of winning. The fact that Bandini was not capable of winning the race, other than by luck and the misfortune of others, probably did not enter Dragoni's head; it was simply that he could see an opportunity to push his protégé forward.

John led the race for thirteen laps and then the differential broke up and he was forced to retire, leaving victory to a bright young newcomer, Jackie Stewart, in a BRM. Bandini did a good job of work and finished second, which enabled Dragoni to report back to Enzo Ferrari that the team's honour had been saved by the brilliance of the young Milanese driving the old car with a smaller capacity engine, while "nostra Campione" broke the three-liter car while in the lead. A statement of fact, but not exactly the complete story. If John had had his way it would almost certainly have been a victory for Ferrari. On the question of John's fitness, Dragoni was still able to say in all honesty "we just do not know; thirteen laps of Monaco was no proof."

The next race was one of the best that John Surtees ever drove. It was the Belgian Grand Prix on the fast Spa-Francorchamps circuit,

283

where the twelve cylinder Ferrari really came into its own. The race was run under appalling weather conditions, the rain starting at the far end of the circuit just as the race got under way. It was a situation that called for a lot of common sense and prudence, as well as knowledge of the circuit and local conditions, all of which Surtees had in plenty. His driving on the wet and slippery track was brilliant, and he drove the powerful Ferrari with all the delicacy of touch and finesse of control that he had developed when racing the MV-Augusta motorcycles. No one could approach him. There were not many people who doubted that John Surtees was one hundred percent back on form, and those who remembered Dragoni's previous suggestions that Surtees was not fit and would never regain his previous form were given food for thought. The Le Mans twenty-four-hour race was imminent and the usual thing was for drivers to do three-hour shifts, with two drivers to a car. After some initial sorties Ford was going all out for victory, fielding a devastating team of 7-liter cars, with a driver line-up that promised to tax Ferrari's resources to the limit. Even before practice began, Dragoni was putting the word round amongst the French press that the Scuderia Ferrari did not have much hope against the might of Ford, and casting doubts on John Surtees' ability to last out the race, especially the three-hour shifts during the hours of darkness and in the early morning when conditions are at their worst.

Although he did not say as much, Dragoni implied that it was because Surtees was not really fit that Ferrari could not hold out much hope for victory. This attitude was now beginning to get John down, and apart from the doubts about his fitness to do Le Mans, he really flew off the handle at Dragoni's pessimistic outlook. As with all the races he took part in, John was at Le Mans to win for Ferrari, anything else never entered his mind, and he felt the whole team should be thinking the same way. After a long and protracted argument with Dragoni he left Le Mans before the race began, going to Modena to see Enzo Ferrari and terminate his contract. The continual undercurrent that had been going on all season had finally broken his stamina, and he could see no future in staying with the team while Eugenio Dragoni was running it.

Enzo Ferrari was very understanding and appreciated that John was disturbed and distressed by the situation, but to him the whole affair was one-sided and perhaps more the result of English and Italian temperaments clashing than anything serious. Dragoni, to all intents and purposes, was doing a good job, running the team according to Ferrari's ideas and getting results, so there was really no justification for considering getting rid of him. The last thing Ferrari wanted was to lose John Surtees the driver, for he was winning races in Grand Prix and Sports cars, which was all Ferrari ever asked of any of his drivers.

On two wheels again. Brands Hatch, 1965.

But if there was friction in the team, and John's driving was going to be affected, then he realized he would have to agree to the termination of the contract. This was agreed and the two parted company the best of friends, both unhappy that things had had to end this way, but both appreciating that there was no alternative. "I often disagreed violently with the team, though Mr. Ferrari and I always saw eye to eye," John said after the break-up.

It was mid-season 1966 when John Surtees reluctantly left Modena and returned to England to drive for the Cooper-Maserati team. He had always been strong-willed and determined, a man of his word and dedicated to any job he undertook. There was no way he could continue with the Ferrari team while the open hostilities between himself and Dragoni existed. The old villain from Milan did not weep any tears at the parting; he was secretly very pleased, for it now meant that his protégé Bandini would be teamleader in the Grand Prix cars. Dragoni still fostered the idea of an Italian World Champion.

However, when he died in 1974 at the age of sixty-five, his dream of an Italian World Champion had still not come to pass.

John Surtees continued to take part in Grand Prix races, first with Cooper and then with the Japanese Honda team, and his supporters in Italy never deserted him. When he won the Italian Grand Prix for the Japanese team in 1967 the crowds went wild. To them, "Il Grande John" was still the same, no matter what car he was driving. Eventually he moved into the realm of racing car constructor, building his own Surtees Formula One and Formula Two cars, which was something that he had long wanted to do. He knew that while he stayed with the Ferrari team he was never going to get into the car building side of the firm, so that a long-term contract with Ferrari had never really been in his plans. Knowing that one day he would have to retire from actual driving he felt his future lay in constructing his own cars, so that terminating his Italian contract precipitated his move

Surtees finished 3rd at the 1965 British G

286

towards becoming a constructor.

In passing, it is worth sparing a thought as to what would have happened if John Surtees had been able to swallow his pride and feelings and stayed with the team, enduring the continual animosity from Dragoni and those who sided with him. Without question he would have won the 1966 World Championship, for he could hardly have failed to win the French Grand Prix, the Dutch Grand Prix and the Italian Grand Prix with the V-12 Ferrari, and as he won the Mexican Grand Prix anyway, with a Cooper-Maserati, it is reasonable to suppose that he could have won that as well with the Ferrari. The car was more than the equal of the opposition and Parkes and Scarfiotti, who were not natural Grand Prix drivers, did well with it. Parkes was second in the French Grand Prix at Reims, and Scarfiotti won the Italian Grand Prix at Monza. Dragoni's hopes for a World Champion from Milan came to nought, for though Bandini was a good lad, he was not in the top rank of drivers and scored only two sixth places by the end of the season.

It is not beyond the realm of possibility that Surtees could have won the Championship again in 1967, for though the trend-setting Lotus-Cosworth V-8 Tipo 49 burst upon the scene and annihilated the opposition while it ran, it did not always run for long, and Denny Hulme won the Championship with a Repco-Brabham by consistency rather than victories. Of the four drivers who had been vying for the crown that Stirling Moss vacated, Jim Clark forged ahead, but John Surtees was undoubtedly in the next position when he left Ferrari. He would have stayed there had he not floundered about with the Cooper-Maserati and the Honda, spending more time worrying about the design and development of the cars than actually racing them. Graham Hill did not make any further improvement to his position, and joining Team Lotus with the new Type 49 saw him have as troubled a season as Jim Clark had. Dan Gurney, the fourth of the likely lads, got himself involved with his Eagle car project which detracted from his driving prowess. So John Surtees, with the engineering and development of Ferrari behind him, and concentrating solely on driving and winning races, would have been very hard to beat.

It is unlikely that John would have been content to remain purely a Grand Prix driver, winning races until his retirement, for the mechanical side of racing cars was fascinating him more and more. While his business instincts were good and he could have won sufficient money to invest for an early retirement, this was not the future he had in mind. What he is doing now was always his aim, and another two years with Ferrari would merely have retarded the inevitable. But my personal feeling is that John Surtees could have had three World Championships to his name rather than one. ✥

Quitting: Surtees leaves the factory accompanied by Eoin Young.

The Later Years— 1961-1975

Although the number of coachbuilders working for Ferrari had been dramatically diminished by the mid-Sixties, the variety of models was enhanced by the appearance of rear- and mid-engine cars. These, together with one or two front-engine models, have been highlights of the more recent Ferraris, many of which have been characterized by conservative good taste but lack the individuality of the earlier pre-production era cars.

Just prior to Pininfarina's attaining dominance as Ferrari's sole coach designer, Carrozzeria Scaglietti—in conjunction with factory engineers—created what may be the loveliest Ferrari of all, the taut and flowing GTO. This model—the last front-engine Ferrari intended primarily for racing—will be found here in both its Series I and Pininfarina-inspired Series II versions.

One of Ferrari's most aggressive but functional touring cars also appeared during this period, the curvaceous GTB designed by Pininfarina and built by Scaglietti. And, for sheer visual excitement, there can be little doubt that the rakish and voluptuous LM has few equals. When the time came for the mid-engine concept to be tried in a road-going Ferrari, the result—the Dino—was a svelte expression of modernity which appears to be wearing very well indeed.

It is really only in comparison with these very special examples that the road-going Ferraris of the past decade seem rather uninspired. Even so, the touring cars remain sophisticated, occasionally lovely, and most important, perfectly equipped to accomplish their intended mission—to transport people at high speed enjoyably, comfortably and safely. In these terms, they have few equals. Other cars may go as fast or handle as well, but there is still only one marque which bears that very special badge and aura. Something of that aura will be conveyed by the portraits on the following pages. (Ed.)

1961 250 GT California with coachwork by Scaglietti ● Owner: J.G. Bennett

1962 Superamerica with coachwork by Pininfarina ● Owner: Gary Wasserman

'62 250 GT with coachwork by Bertone ● Owner: Bill Karp

1963 GTO Series I with coachwork by Scaglietti ● Owner: Paul Pappalardo

1963 250 GT SWB Berlinetta with coachwork by Scaglietti • Owner: Jim T

330 GT 2+2 with coachwork by Pininfarina ● Carrozzeria Pininfarina

1964 Superfast with coachwork by Scaglietti ● Owner: Pete Ci...

3 250 P with coachwork by Fantuzzi ● Owner: Don Fong

1964 250/275 LM with coachwork by Scaglietti ● Owner: Robert N. Dusek

1964 250 GTO Series II with coachwork by Scaglietti • Owner: Carle C. C

1965 Dino 206 GT prototype • Owner: Carle C. Conway

1965 275 GTS • Owner: P. H. Knowles

1966 365 P • Owner: G. E. Keeney

1967 365 California ● Owner: Burt Born

Frank Weinberg

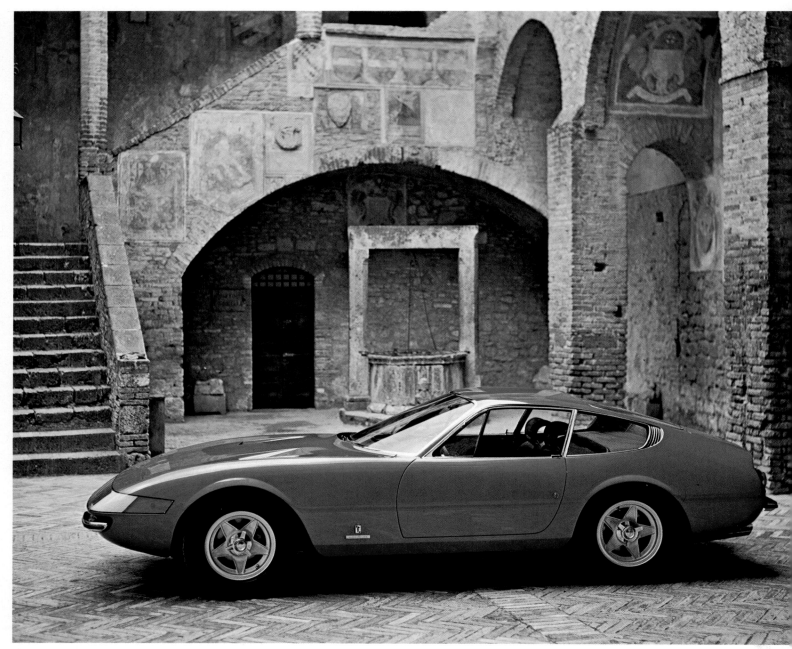

1968 365 GTB/4 Daytona ● Carrozzeria Pininfarina

1969 365 GTC ● Owner: John Knowles

1969 365 GT 2+2 ● Owner: Guy G. Ang

73 Dino 246 GTS • Owner: Gil Wiener

1973 GTC/4 • Owner: Gil Wie

365 Berlinetta Boxer • Maranello Concessionaires Ltd

1975 Dino 308 GT/4 2+2 • Maranello Concessionaires Ltd.

1975 365 GT/4 2+2 ● Maranello Concessionaires ▶

Photo
Credits
for this
Section

289: Photograph by Henry Austin Clark Jr. 290, 291, 296, 297, 303, 304, 309, 310: Photographs by Marc Madow. 292-293, 298-299, 300, 301, 305: Photographs by Stan Grayson. 294: Photograph by Rick Lenz. 295, 306: Courtesy of Pininfarina. 302, 307, 308, 311, 312: Photographs by Neill Bruce.

NOTE: three captioning errors were discovered after production of this color section was completed. The Superfast on page 296 carries Pininfarina coachwork, not Scaglietti. The captions for the 308 Dino on the page opposite and the 365 GT/4, page 311 below, were reversed.

CHAPTER 13

Today and Tomorrow–
Ferrari & Fiat

by Griffith Borgeson

It seems safe to say that the automobile industry is today in one of its most precarious positions since the Great Depression. Oil prices, speed limits–even in Italy, ever stricter emissions controls, economies sagging under inflation, all these have raised question marks about the auto industry's future. In such times, there is perhaps no segment of the industry which seems so visible as that composed of builders of superfast, superexpensive machines like Ferrari. What does the future hold for Ferrari? Griff Borgeson traveled to Italy to have a firsthand look at both the company and at a country about which he cares deeply.

On the way to Maranello from the south of France I rolled into Turin in mid-January of 1975. I had been away from Italy for two years but had kept up with Italian news in the foreign media. On the basis of that intelligence I expected a country in chaos and on the brink of revolution. One good look at Turin made it overwhelmingly clear that while there might be something of a crunch going on—as in a good many other parts of the world—conditions nevertheless were extremely normal. As a matter of fact, the great automotive capitol seemed to be more affluent than ever. This was disorienting and, in fact, incapacitating. I could not take hopefully valid readings on the future with a fictionalized view of the present as a point of departure.

Gianni Agnelli.

I went immediately to Luigi Giovanetti, the secretary-general of the ANFIA, Italy's powerful and authoritative Automobile Manufacturers' Association. He confirmed, with considerable intensity and outrage, that a grossly distorted image of affairs in Italy was being projected by her friendly competitors for world trade. He had piles of documentation of the sort that I had been seeing and, in fact, the January issue of *Auto 70* carried a feature article by him on this very subject.

"Whether it's deliberate," Giovanetti said, "or just a result of unprecedented ignorance afflicting professional journalists, the negative result is the same: we're being made to look infinitely worse off than we are."

I felt that I had a fairly good idea why this bad-mouthing should have attained such an unusually shrill pitch. One, the competition for markets had become desperate. Two, the swing of the Italian government all the way to the Left had been anticipated for well over a decade. Perhaps that time was drawing near and Italy's traditional partners knew it.

I called my highest contact in the industry, a man much more highly placed than Giovanetti.

"What," I asked him, "are the prospects for a takeover of the government by the Left?"

"Forget about 'prospects,' " he said. "It's a matter of inevitability and it's not too far off. I'm no Leftist, God knows, but things may run better then. Let's hope so."

What surprised me was his air of certainty. Because of their solid reputation for honesty in government, the Communists were attracting ever larger segments of the electorate. But if Portugal should go the way of Hungary they would be finished in Europe for a long, long time. In any case, it was time to find out what the Communists thought.

I contacted government Deputy Lucio Libertini, long-time Socialist, now a Communist, former vice-president of the Chamber of Deputies' Commission on Industry, and a noted theoretician on industrial planning on the national scale.

"Are the Communists finally going to take over?" I asked him.

"No. We don't have the numbers. The Left stands a certain chance if we, the Socialists, and the Left of the incumbent Christian Democratic Party can form an effective coalition. That's where it all stands at present."

Giuseppe Dondo in 1973.

"And if the coalition gets elected, what will happen to Italian industry—to Fiat, for example?"

"All the *big* productive entities which are fundamental to the economic life of the country, of course, will be nationalized. They must be converted from serving the interests of a small elite to serving those of all the people. Otherwise, business and private property will go on essentially as usual."

"Ferrari, being a part of Fiat, would be nationalized then?"

"No doubt."

"Can you see luxury GT cars being manufactured in such a Workers' State?"

"Certainly. To support our economy we will have to do like everyone else, that is, produce, sell, and export those things which we are particularly good at making. Ferrari is a good example of that sort of thing."

"And what do you think of the prospects for Ferrari continuing to race under a government of the Left?"

"At the end of World War II Italy was in ruins and was one of the most poverty-stricken countries in the world. She rose to be the seventh richest country in the world. Throughout that whole period, and from the very beginning, Ferrari's racing efforts have broadcast and enhanced the prestige of Italian industry throughout the world. We may need that even more in the future. It is hard to conceive of throwing that away, or of it really being incompatible with the goals of a truly popular government. These are relatively small considerations and they await study; I can only give you my opinion."

I spent seventy-two hours in Italy's industrial capital, discussing the crisis and the future with the broadest possible cross-section of the population. I arrived at the feeling that the traditional system was at the end of its natural life-cycle; that it was for this reason that change was "inevitable"; that the pangs of death and rebirth are difficult but not tragic; and that the transition was taking place in, on the whole, an admirably orderly and democratic fashion. For the mass of the population, private property would not change, nor would the mechanical genius of the people, nor would its passion for divine machines.

With these thoughts in mind I pushed on to Maranello, to spend an afternoon with Ing. Giuseppe Dondo, Fiat's top man there. Fiat had given me a brand new Tipo 132 to savor on the trip. The car was wonderfully smooth, solid, and responsive; just another one of those

Carrozzeria Scaglietti, Modena.

Dino 246 in assembly jig.

Fitting a Daytona rear window frame.

Dino assembly.

At Scaglietti, body shells await painting.

Seat assembly.

things that the Italians do surpassingly well. As it sung over the magnificent autostrada I reflected upon the interrelationships of Ferrari and Fiat destiny, going back first to Enzo's talent-raid on Turin in 1923, to subvert Vittorio Jano. And under the hood of the 132 hummed a splendid engine which bore the stamp of Aurelio Lampredi, who had won his professional spurs at and with Ferrari. The car's porcelain-like finish was of that deep midnight blue which had been the stamp of the cars built by Vincenzo Lancia—the man who, wrestling a big Fiat racer at Bologna in 1908, had fired the ambition of the ten-year-old Enzo to seek a career in the thrilling new sport of motor racing. The past was a single harmonious tapestry, if only one could learn to recognize its threads . . . threads from which the future will also be woven.

Along with the good and orderly citizens, I held to the legal 120 kph (75 mph) speed limit, thinking, "Who needs a Ferrari now? An epoch has ended." At 333 kilometers from Turin the fuel gauge was very low. I stopped at one of the rare service stations and asked for a full tank. It gulped down 39.7 liters of Super at 300 lire per—$18.35 to cover 207 miles in an 1800 cc car. As I peeled off the banknotes I wondered what the tab would have been for the joy of driving a Ferrari. Yet, at these numbing prices, Italian traffic was so dense that cities such as Rome and Naples had become paralyzed to the point that they were being forced to blockade large urban areas against the

insupportable invasion. A "poor country" indeed!

I was taken aback to find the factory gate at Maranello wide open and unguarded; I simply drove in and parked in the courtyard. Long-time public relations director Franco Gozzi came out to meet me with a warm greeting and said, "Ing. Dondo is looking forward to your visit. Shall we go in?" This was a far cry from kicking the snows of yesteryear.

Giuseppe Dondo is a tall, slender, sandy-haired man who was born in 1933 and who brings youthful drive and professional discipline to the post of general manager of what is now Ferrari S.p.A. (S.p.A. = Societa per Azioni = a shareholders' corporation.) He took his engineering degree at the Turin Polytechnic University in 1958 and moved directly into the Fiat factory system. He did a tour of duty for his employer in the United States and speaks fluent English and French, in addition to his native tongue. He has represented the Fiat presence at La Ferrari since the Ferrari-Fiat accord of 1969. It was that accord and its consequences which had brought me to Dondo's strictly simple and businesslike office in Maranello.

On Dondo's desk, for handy reference, reposed a copy of a very large book, bound in red cloth and embossed with the single word, *Ferrari*. It was the seventh Italian edition (a very limited edition, not for sale and not available to the public) of Enzo's autobiography, richly improved in its graphic content and with numerous changes to

depth of my feelings as never before. Then Agnelli spoke. Younger than me by over twenty years, I felt in him the force of the modern man, of the politician and the diplomat of the business world, of the vital and all-perceiving observer. The questions he put to me were concise and precisely to the point, those of the man who wants only to know and to understand. At the end he called in his aides and concluded, 'Well, Ferrari, couldn't we have done this a long time ago?' And then, turning to his own men, he said, 'Gentlemen, perhaps we have lost time. Now we have to regain it.' ''

"That's all there is," said Gozzi. "We, the Ferrari aides, and Avv. Agnelli's men sat in an anteroom for an hour and a half while our chiefs had their meeting. There were no witnesses to what took place."

If no one aside from Ferrari and Agnelli knows the slightest thing about the financial aspects of the accord it's no one else's business anyway. As for the organization of the reconstituted company, all seems to be very clear. One, exactly fifty percent of the stock is owned by Fiat. Two, Enzo Ferrari is the president of the company, backed up by its Fiat-appointed general manager (*direttore generale*) and managing director (*amministratore delegato*). Three, the company is separated into two distinct divisions, Racing and Industrial. Enzo Ferrari is officially in charge of and responsible for all activities of the *Riparto Corse* and Fiat, through its two board members, is entirely responsible for the *Riparto Industriale*, the GT-car division.

This is the formal distribution of responsibility for legal purposes but, in practice, there is a steady flow of cooperation between the two divisions. All major decisions affecting the company as a whole are arrived at jointly by the three-man board. An obvious example of this pooling of authority would be the framing and adoption of a plan of action for each new business year. Then, the GT division depends very heavily upon Enzo Ferrari for leadership and guidance, for reasons which Ing. Dondo makes clear in defining the nature of his job:

"Although sent here by Fiat, I try, above all and at all times, to be a man of Ferrari. I try to comprehend the spirit of that which Ferrari cars represent, the spirit which our clientele expects from a product which bears the Ferrari name. I try to understand the mystique of what Ferrari means in the world, to safeguard it and to develop it in the future. I try to define these criteria, both tangible and intangible, and to concretize them. In many ways it is almost a metaphysical exercise, and who can be of greater help in coming to grips with it than the creator of the phenomenon himself?"

The third board member, the managing director, is Francesco Bellicardi, who also has been for many years the general manager of the Fiat-owned Weber carburetor company in nearby Bologna.

bring its text up to date. Earlier editions of the work had appeared in Belgium, Czechoslovakia, England, France, Germany, Holland, Japan, Portugal, Spain, and the United States, in addition to Italy.

I began our interview by saying that the Ferrari-Fiat accord was an event of great historic importance—certainly to all lovers of the Ferrari marque—and that almost nothing concerning it ever had been made known publicly. I hoped to learn more about this mysterious arrangement.

Dondo thumped the massive red book and said, "The only and sole historical reference that we have pertaining to that subject consists of Ing. Ferrari's own words, as recorded in his autobiography. The famous *affare Ferrari-Fiat* was an understanding arrived at privately between Ing. Ferrari and Avv. Agnelli (Avv. = Avvocato: attorney). Since Avv. Agnelli has made no comment on the matter we are left with Ing. Ferrari's observations."

What Ferrari chose to say was that on June 18th, 1969 he found himself sitting before Gianni Agnelli in his inner sanctum on the eighth floor of Corso Marconi 10 in Turin. And then, as Ferrari describes the meeting:

'' 'Ferrari, I am here to listen to you,' [Agnelli] said, speaking with the secure elegance of the president of Fiat. I spoke for a long time, and he did not interrupt me. I told him of my yesterday, my today, of the tomorrow of the factory, and I was able to express the

Steel billets for V-12 crankshafts.

Machining a Dino crank.

V-12 assembly.

Dino assembly.

In the old days of the Scuderia Ferrari, Enzo knew Edoardo Weber well and worked closely with him. After the war and Weber's death, when Fiat acquired that company, veteran employee Bellicardi was put in charge. He is not a great deal younger than Enzo and approaches his work with a sort of joyous seriousness which is very similar to Ferrari's. The two men have worked closely together for decades and, with so much in common, are very good friends. Bellicardi's appointment as a director of La Ferrari at the time of the 1969 accord was a beautiful stroke of corporate diplomacy in that he was already practically a member of the family of each of the parties to the accord. Bellicardi continues to have his headquarters at Weber in Bologna but is in constant contact with his colleagues at Maranello, which he visits regularly.

Thus Enzo Ferrari, as president, heads a troika management of the company which he founded, and of which he remains the vital inspirational force and symbol. While being, technically, one hundred percent in charge of the Racing division, he is highly attentive to the counsel of his co-directors, who represent the interests of his 50/50 partner, Fiat. And they, technically one hundred percent responsible for and in charge of the GT division, would make no important decision without seeking Enzo's opinion.

Financially, the GT division functions entirely on Fiat capital. The Racing division, of course, derives a considerable portion of its annual budget from outside sources, such as appearance money, prize money, and subscriptions from sponsors. In addition to these funds, Fiat helps to underwrite this division's operating costs. How the totality of these external and internal funds is spent is entirely in the hands of Enzo Ferrari. Finally, it is the sum of the balance sheets of the GT and the Racing divisions which constitutes the total balance sheet for the firm as a whole.

From Enzo's point of view, he can go on racing for the rest of his life with the absolute minimum of financial worry and without the distraction of having to run the GT division in order to stay alive. Fiat gets all the assets and goodwill of the world's most distinguished and prestigeous GT marque, plus the continuing leadership of its creator, plus all the intangible benefits of continuation of the racing activity which has been fundamental to the marque image.

Ferrari's quest for some such solution to keeping himself and Italy in the forefront of international motor racing began in 1934, in response to the new German domination of the sport. He developed and promoted the idea of a Scuderia Italia, to be supported by Italian industry as a whole. His considerable talents as an organizer and persuader were unable to overcome corporate individualism and jealousy and nothing came of these efforts. After the war he took matters into his own hands, with the phenomenal success which the whole world knows. As the forces of history massed against his single-handed struggle in the early 1960's, he revived his old plan and tried once more to form an industry-financed national racing organization. This time his credentials were of the most compelling, but the effort failed again. Then in 1964, with both the Ford Motor Company and the FIA against him, Ferrari told his friend Bellicardi of the FIA regulations which had been announced for the 1965 season. Their requirement that the power plants of Formula Two machines must be made in quantities of no less than 500 units per year meant that Ferrari would have to abandon this level of competition, important in itself and important as a development facility for Formula One. It was Bellicardi who conveyed this news to Agnelli, with the result that Fiat opted to produce the original Dino engine in volume and to build the Fiat Dino GT around it. Fiat seems to have given no other significant support to Ferrari's racing program, but this very important gesture had other consequences and paved the way for the accord of 196

The bodies for the Dino Coupe were built for Fiat by Bertone. Since the mid-1950's, Ferrari GT coachwork had borne the Pininfarina

Boxer prototype at Fiorano test track.

emblem so consistently that it was generally taken for granted that an exclusive contractual arrangement existed between Ferrari and PF. This, according to Dondo and Gozzi, never was the case. It was simply a practice which was established on a foundation of mutual respect between Enzo Ferrari and Battista Pininfarina—respect which grew into a firm friendship.

The Fiat Dino engines were made in Maranello. Toward the end of the Dino Coupe program Fiat elected to have Bertone deliver the bodies to Maranello and to have complete assembly of the cars carried out in the Ferrari plant. In this way Ferrari executives became impressed by the originality of design concept in that vehicle and also got in the habit of working with key Bertone personnel. This led Ferrari to decide to see what Bertone might come up with as a design for a more purely and aggressively Ferrari type of car. This resulted in the Ferrari Dino 308 GT4, introduced at the Paris Salon in September of 1973, and in Bertone sharing with Pininfarina the honor of being coachbuilder to the house of the rampant colt.

Over the decades, the bodywork for Ferrari racing cars has been the specialty of Carrozzeria Scaglietti of Modena, an artisan shop of modest size and known for its skill in the hand-forming of aluminum sheet. The designs for Scaglietti-bodied Ferraris have been narrowed between body design for Sports Prototype and GT cars suitable for either street or track, and as the trend moved toward monocoque construction for competition cars of all types, the latter came to be built in their entirety on the Ferrari premises. And, while this was taking place, Scaglietti converted to meet the demand for Ferrari's more far-out forms of GT bolide, always designed by Pininfarina. Following the accord of 1969, and foreseeing an enduring place for strictly hand-built models in its product line, Ferrari S.p.A. acquired Carrozzeria Scaglietti as a subsidiary. Bellicardi is its president and Dondo its general manager.

Plans for the future of the GT division are to continue to design and build the world's most advanced and perfect road machines in the classic Ferrari tradition. The market for such cars has been touched only lightly by recent world events, including inflation and the petroleum crisis. Higher gasoline prices have not proved to be a sales

365 GT/4
Boxer Berlinetta

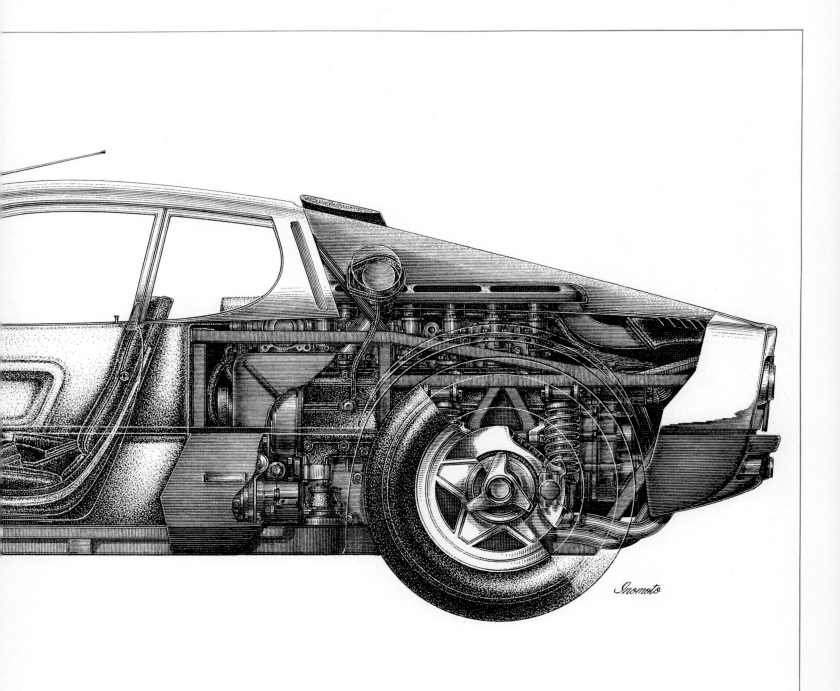

Inomoto

Boxer prototype at Fiorano test track.

deterrent on the upper levels of the luxury-car market, nor have price markups due to inflation particularly disturbed the buying habits of this clientele. The imposition of relatively low speed limits throughout Europe, on the other hand, has had a marked effect upon the sales of fast, powerful cars on the Continent. These laws without precedent came into effect with a suddenness which left European motorists stunned. Ing. Dondo takes courage in the example of the United States, where moderate speed limits are virtually as old as the automobile, yet where Ferrari enjoys one of its best markets. He hopes that Europeans soon will make the same adjustment to these restrictions which Americans made long ago, and learn to value the fine GT car in terms of its other manifest virtues: superior braking, steering, roadholding, reserve of emergency power, and mechanical excellence in general.

The Fiat administration at Maranello has brought new organizational and marketing benefits to Ferrari and, while European sales have registered a temporary lag, exports have increased and go on doing so with the ripening of new markets. Thus the old Ferrari production ceiling of 1,000 GT cars per year has now been raised to about twice that figure. The plant's potential productive capacity is about 2,500 units annually, and greater volume will reduce unit cost, favor the price-competitiveness of the product, and provide additional funds for keeping the little artisan industry in the forefront of technological development in the pure Ferrari tradition.

Probably the most burning question concerning the future of the marque is, "Will Ferrari, under Fiat, continue to race?" At this moment in history, with so much in ferment, firm forecasts are impossible to make. But Ing. Dondo does not hesitate to express his clear-cut opinions:

"Today, La Ferrari is known in all the world as *the* great Italian racing marque. When Italy was destroyed by war it was La Ferrari which went into battle to raise our hope, our self-confidence, and the prestige of our industry. Today the name is known in the most obscure corners of the world and, throughout the world, it has served as a banner for Italian industry. It represents a great treasure of experience, of talent, of know-how, of dedication, of spirit. Even when, in racing, it finds itself in the position of the underdog, it still is a symbol of courage and of the will to overcome terrible odds. Fiat, in making its accord with Ferrari, also made a commitment to racing. I find it inconceivable that La Ferrari should cease racing. If that happens, there is no racing left. It would be the end of a symbol and the end of an epoch. Ferrari cannot abandon racing."

I thanked Ing. Dondo for having told me those things concerning the present and future that I had come to Maranello to learn. As we left his office Gozzi said, "Ing. Ferrari would like to greet you before you leave. It will only take a moment." ❖

CHAPTER 14

An Audience with the King

by Griffith Borgeson

Interviews with Enzo Ferrari are notoriously rare. Of those which have been published, few have not reflected tension on the part of both the journalist involved and Ferrari himself. Because Ferrari is the man he is, interviews with him have seldom been enlightening, too often seeming staged or guarded, as doubtless they were. Few if any journalists have been granted the opportunity awarded Griff Borgeson who–at Ferrari's invitation–spent nearly two hours talking about a wide range of subjects with the Commendatore. Mr. Borgeson made the most of this time and what follows probably represents the most candid, revealing interview with Enzo Ferrari ever published.

I had come with no expectation of seeing Enzo Ferrari. We had met a couple of dozen times over the years, but always when each of us was overwhelmed by great demands upon his time. He had always been politely cordial, never accessible. I had been content to learn about him through his writings, those of others, and through observing him at a distance.

I had never been in his private office. The high-ceilinged room is about thirteen feet wide and thirty feet long. We entered at one end and there was the old King, at the far end of half an acre of leaf-green carpet. For three-quarters of their height the walls were tapestried in textured cloth of a tint almost as dark as the rug. The baseboards and picture molding seemed to be in red mahogany and the indirectly lighted ceiling was cream colored. The only furniture that I noted in

Monza tests, 1959.

the whole dimly lit expanse consisted of a desk and three chairs, also in natural wood. The green walls also were almost bare, except for a large photo portrait of Signora Laura, a painting of a hurtling red racing car, and a large color print of Francesco Baracca, Knight of the Air, standing smartly beside his Spad biplane.

Ferrari, seated at his desk, did not look up. He was engrossed in impressing his fine script upon a notebook, using the same bright purple ink with which he had signed the first letter I had received from him a quarter-century ago. There was something hauntingly familiar about the scene; then I remembered an impression of Bonaparte, utterly alone, working at his portable desk in full military campaign, in a forest clearing. Gozzi and I stood before the desk in silence for at least a full minute. Then Ferrari looked up, expressed delighted surprise at our presence, and said,

"Borgeson! Buon Giorno! Piacere!"

He rose to his full height and took my hand in a strong grip. He looked splendid. He showed his seventy-six years but bore them bouyantly, it seemed. His hair had been snow-white for a long time. His skin had always had that Po Valley pallor. His jowls were looser but the lines of his face were good—relaxed and showing no inner tension. His massive shoulders were not as square as one could remember them, but he still moved with a great deal of suppleness. It was warming to see a veteran of so many battles so unscarred. At his invitation we occupied what remained of the room's furniture and began to talk. Ferrari wanted to know if there was anything that I would like to ask him.

Had I been prepared for such an opportunity I certainly would have spoiled it with a boring list of minutiae about his career. As it was, we merely chatted.

I had had good reason for wanting to know his father's name, and I asked it.

"My father was named Alfredo," he said. He was from Carpi, not far from here, and he was the son of a *salumiere*—a man in the delicatessen trade. He started out as a technical employee in a railway-equipment factory in Modena. Then he set up that little shop with ten or eleven employees, where I was born."

I had assumed that Alfredo-Alfredino-Dino had been the traditional name of firstborn males in this branch of the Ferrari family for generations, and this was some confirmation. In erecting monuments such as cars and schools in the name of Dino Ferrari he was certainly honoring his son. But, consciously or not, he was at the same stroke honoring his brother Dino and his father Dino and perhaps a long line of forebears of this name.

I also asked the name of Enzo's mother and learned that it was Adalgisa Bisdini. She came from Forlì in Romagna.

Then I asked this man of surpassing talent and power and of more than ordinary erudition what, precisely, his formal schooling had consisted of.

Modena testing 1960, Enzo and Chiti.

Modena testing 1960, Enzo and Chiti.

1955 Super Squalo.

"Not much," he said. "The first four years of elementary school and then three years of trade school. That was all."

"But you went on to steep yourself in Stendahl, d'Annunzio and . . ."

"D'Annunzio!" he said, "Of course he was irresistible to me in my youth. Not only was he a great poet and author, he was mad about automobiles. He had splendid Lancias, and then he began to buy Alfa Romeos, when I worked there. I began with his *Il Piacere* and *Il Fuoco,* went on through all his works, then read all his critics and biographers. But I mentioned him and Stendahl in my book only in passing. Actually, my literary interest ranges from Kafka to Einstein and a bit further."

Ferrari spoke, leaning toward me, elbows on desk, and with his eyes boring into mine. He expressed himself in superbly modulated Italian—as a trained orator or opera singer might do. At the same time, his voice was full of sympathetic warmth and his face expressive, often smiling. And whatever he said he said with an air of finality. One had only to pose the questions. He had lived long enough to have reflected upon them all and to have his answer for each.

"I read everything," he continued. "I believe that a man should enrich his mind by remaining open to all ideas. It's like music. I can enjoy a lyric opera, but also a folksong, light dinner music, an operetta. They all penetrate my ear easily, and it's the same with literature. I read all that my education or intelligence permits me to absorb, and I put aside those things that I find too difficult to understand. Since boyhood I tried to read Benedetto Croce, a great philosopher. I never could be sure that I was interpreting him correctly and I finally gave up. I am an undisciplined reader. I can pass from an erotic magazine to a novel to a gripping tragedy; I have found an infinite sweetness in reading the life of Leopardi. But these are all sensations which happen to harmonize with certain periods of our life. Literary works, I believe, are useful to the state of soul which we have need to confront or to confirm.

"I have read a great deal of Emil Ludwig. I read his *Napoleon* because the personage of Napoleon has always fascinated me. For me, he was a great general, a great mathematician, a great legislator, and also a great lover—and the beauty of woman did not hinder him in his other achievements. And I have scoured the literature on Napoleon, in order to arrive at a concept of the man."

"Your concern with philosophy," I said, "certainly permeates your autobiography."

"If so, it is not because it is the work of a philosopher," Ferrari said. "It was written by a man who draws upon his heart and upon his conscience. In writing I have no need to construct anything, which I find to be difficult anyway. When a man merely writes what he feels the problem is just one of choosing words, adjectives. The adjective that I would use to describe that book of mine is not 'philosophical' but 'spiritual'."

"It's also a bitter book," I put in.

"Look," he said. "I am obviously satisfied with what I have accomplished, even if I have paid for it with all that I ever had. But if you were to ask me today, 'What should we wish for a baby coming into the world?' I would say: That he be born an orphan; that he be born rich; that he be born impotent. An orphan, so as not to have to suffer the loss of his parents and relatives and all the other consequences that go with being part of a family. Rich, because I have known what it is to suffer hunger. Impotent, so as not to take a single hour from his work day for pleasure, but to dedicate himself totally to racing cars!"

These last two words gave comic relief to what had been an entirely serious statement, and we all laughed. Still in that mood I asked, "And have you lost so very much time in your career through the curse of. . .non-impotence?"

He laughed at the *mot juste,* then said, "No! No, I have not lost time. But there is no question that that is always a complicating factor in life. And I have had the great sorrow of losing my son, and so I know what this complication can lead to. But I have lost little time due to women for *one* reason. Because I have always considered woman to be not just a necessary element of our life, but the reward for work. I have never put woman before labor. After the day's work is done, woman is the prize. But not before. Never!"

"I see that, since Fiat's arrival, one sees a few women working here for the first time. Why did they always seem to be taboo in the past?" I asked.

"It's not the first time," Ferrari corrected me. "We had them during the war, too. I am not opposed to having feminine employees, providing they form about half of the work force. It's when you just have a few that they tend to form little cliques and to have a disturbing effect upon the rest of the workers. In any case, I do not like to see women being made to do heavy work. I can accept that they do light work, if they must, but I still prefer to think of them as the prize."

"And about men," I said. "Would you like to say something about the men who have worked with you to create La Ferrari?"

"Not in detail, no," he said. "I would be afraid of forgetting someone. But I would like to say this—that *all* who have worked at La Ferrari, even briefly, have made a meaningful contribution. Why? Because I have always believed that the relationship between the company and a mechanic, a machinist, a designer, a draftsman, an engineer, is a fifty-fifty thing. For the simple reason that when

Laura and friend, 1961.

335

someone comes here he brings with him his experience, his attitudes, and his enthusiasm. When he leaves, he takes with him what he has learned here. But he leaves a part of himself, and this enrichment works in both directions. I try to avoid making individual evaluations, but it is obvious that he who stays longer gives more and that he who stays a shorter time gives less.

"Certainly, one who has given enormously and who is still living is *Cavalier* Bazzi, an example of absolutely formidable dedication. Then Jano. Jano! He was undoubtedly the greatest technical mind that I have known and he, above all, has left the most profound impression upon my sentiments. And we remained close until his death.

"He was a man whom the times never overtook, never outdated. He was a man of ideas, but never of strange ideas. It was almost unheard-of that a machine conceived and developed by him would have a point of weakness, because of the thoroughness of his attention to every minute detail. Even when his conceptions were not daring, the results were always honest and sound. This is to speak of just one man, but so many have passed by here. It is impossible to evaluate their relative merits.

"For me, too, the occasion of a man's death is not the time for evaluating his worth. It is in that circumstance of a man's death that journalists rush into writing all that they have never said nor even thought about the person during his active life. What counts for me, instead, is that a person, throughout his active life, should maintain a certain esteem for his work in the eyes of his fellow men. The secret is there."

Ferrari inquired about my session with Ing. Dondo and I told him of our discussion of the future of La Ferrari.

"The future," he said, "is foreseeable in moments of normalcy. But today we live in an exceptional moment. Suppose that I ask you, 'How and when will this energy crisis be resolved?' I believe that there is no one in the world today who can say that it will last three months, a year, two years. The future of La Ferrari, in a period of normalcy, such as a year or two ago, was fairly foreseeable, even after my time. For the simple reason that it is a technological patrimony which should not and cannot disappear. Why? Because I have always had the conviction that factories are made, first, of men, then of machinery, and then of walls. Here, if we have nothing else, we have a very, very important human patrimony.

"I contend that a good 65 to 70 per cent of the worth of any business concern exists in its human wealth. Now, the dispersal of such a patrimony certainly would not be the will of Fiat, but a result of the conditions of the moment. We cannot know what conditions may be in six months or a year. At least I cannot conceive of what they will be. I say to myself, 'We have lived all this before.' I

remember the crises caused by the war in Ethiopia; the Mille Miglia we ran on an improvised 'national' fuel: the crises caused by World War II; the closing of the Suez Canal in '56. But now we have a situation in which those conditions have been dilated macroscopically and we have cyclopean forces on our hands. They are not controllable, and any forecast of them can only be a guess, not a conviction.''

I asked Ferrari if he was familiar with the projections of the Club of Rome for the future of technological civilization.

''Yes,'' he said. ''The growth factor inevitably leading to catastrophe. It is enough that, among the persons who exercise world power, there be one madman—as we have known to happen in other circumstances—for all equilibrium to vanish overnight. What and who is there to restrain the ambition of man? Nothing, no one. Each man is a world unto himself, with his anguish, his torments, his infirmities. And when infirmity touches the brain of a truly powerful man, the social equilibrium which saner men have managed to create is finished.''

We spoke of Italy's political horizon and the lack of any strong individual leadership.

''The only hope,'' he said, ''is that good sense may triumph and that various men may forge the minimum of accord to keep us from attacking each other. I think that it is tragic for us that the concept of *Patria* has vanished from the written and spoken Italian language. Fatherland means profoundly more than mere country or nation.''

''What future,'' I asked, ''can you imagine for La Ferrari under a Socialist government in Italy?''

''That will depend upon the conception of motor racing which these new political leaders bring with them. Because La Ferrari cannot

exist if racing activity is denied her. Now, since the existing Socialist countries place such great importance upon athletic achievement and competition, it is perfectly possible that automotive sport would continue to have its place here. The Soviets have made a certain place for it in their system, even without having any tradition of the sort. It could go very well. For me, it is clear that racing could become an important element of prestige in such a new government. And until now, the Left has said nothing to suggest that things might be otherwise. We can only wait and see. We may be better off, and maybe worse. It's like leaving a lover. One always expects the new one to be better but often one would like to be able to go back to the old one.''

I asked Ferrari what endowment he felt had been most important to him in his career.

''In this type of work, for me,'' he said, ''the most important single factor is the lightning flash which is the birth of an idea. At some utterly unexpected moment you wake in the middle of the night and say, 'That's it! I've got it!' And at that point you begin to mold that idea in your brain, and carry on until at last an entire project has been achieved in all its detail. It is the birth of the idea which is the most important and the most difficult. . . that, and the task of convincing others that the idea is a valid one. Take Einstein and his Unified Field Theory for example. It took him years merely to find the mathematical language for communicating it to others. Building racing cars is not all that different.

''To persevere in this activity you have always to maintain an ample reserve of illusion, of hope and dreams. You conceive and you build a racing car. And the day that it takes its place on the starting

337

grid you realize that it has so very many defects. This is why I always say that I have yet to think out that most beautiful of cars that I would like to create. Perfection does not exist; only evolution. Constant and necessary evolution. . .and always problems terrible enough to destroy you. That's life, and one has to learn to wait."

My final question was, "Is there a personal philosophic basis for this creation of yours, La Ferrari?"

"La Ferrari," he said very deliberately, "is the living expression of my dreams. To have dreamed machines and to have realized them is a rather beautiful and fascinating thing. It is because of that, that after so many years, I still continue to dream. And there you have it all."

"Thank you," I said, and then quoted, in Italian, these lines by Lawrence of Arabia:

"All men dream; but not equally. Those who dream by night in the dusty recesses of their minds, wake in the day to find that it is vanity; but the dreamers of the day are dangerous men, for they act their dream with open eyes, to make it possible."

"*E tutto lì,*" Ferrari said. "It's all there! And what a man Lawrence was!"

I was preparing to take my leave and we were standing when Ferrari said,

"Eh, Borgeson. There is something else that you should know. As you know, I am the president here, and exactly half the stock is mine. Also, the accord with Fiat is *in vitalizio.*"

Yes, that was very important, *Vitalizio–viager* in French—is a juridical arrangement which is little known in the Anglo-Saxon world, but very commonplace on the European continent. It works this way. I am no longer young and I have accumulated property. I can't take it with me and would like to benefit from it while I live. So I sell it *in vitalizio* and the buyer pays me a fixed income—rent, for example —for the rest of my life. Until my death, *I* remain the owner whereas the purchaser has the use of the property. It is a form of gambling with death. If I die a few months later, the purchaser takes full title and has a fantastic bargain. But if I live for a long time he may pay very dearly for his acquisition. Among other things, *vitalizio* is an incentive for living a long time.

In Ferrari's case, it means that all of the conditions of the 1969 accord apply for the rest of his life. That is to say that his presidency, his fifty per cent control, whatever annuity he receives from the purchase, *and his one hundred per cent control of the racing division* are his for the rest of his years. In a very real way, life began for Enzo Ferrari in 1969.

"That's just beautiful" I said as we parted. "May the next hundred years be filled with wonderful dreams realized."

"Thank you," he said with a bright twinkle in his eye. "But come

back soon, anyway."

I headed back to Turin and spent a thoughtful night. Then, before putting the Alps behind me, I had a final session with an old friend.

"How did it go?" he asked.

"The session with Ing. Dondo was splendid. And, to my great surprise, I was able to spend almost two hours with Enzo Ferrari."

"I haven't seen him for years. Tell me about him. How does his health seem to be, and his state of mind?"

"He seems to be just fine, both physically and emotionally," I said. "He has a serenity which I doubt that he had before. The tension and bitterness which he used to emanate seem to be gone. He smiles and laughs a great deal, as he did in so many of those photos of his youth.

"He now speaks of Dino's loss as a fact of life, and not as his personal flaying by a cruel, blind Fate. I may have gained more insight into that situation. Of course parental love was a great and real factor. But I think that a perhaps equally great one was the father's practical need for his son at that time. He needed a strong young warrior to stand by his side and to gradually assume the full burden of leadership in the struggle, *l'opera,* the Great Work—La Ferrari. Only Dino could have filled that role, and he had been groomed for it from childhood. Then, just as Dino was reaching maturity and Enzo was beginning to feel his years, he was stripped of that one possibility. He was condemned to abandon the struggle and the dream or to climb back into his armor and to carry on with declining powers and utterly alone. So of course he railed wildly against the gods.

"By grace of his really astonishing courage, will, energy, intelligence, and cunning, he managed to *win* his Waterloo, and to do so at the eleventh hour. He legitimized his crown, consolidated his kingdom, and executed the diplomacy that would enable it to endure. His reign is assured for the rest of his life and his life's struggle is, very simply, won. And so he is beautifully serene although, I'm sure, still a tiger."

"Griffith," my friend said, "what you say makes me deeply happy. But there is one more thing that you should know. It was Ferrari who came to Gianni Agnelli in 1969, and not the other way around. I am *sure* that, at that point, Ferrari was absolutely against the wall, with no prospect other than bankruptcy and the ruin of all that he had ever worked for. Gianni is a *dur Piemôntais,* just as hard as his grandfather, the old Senator Agnelli, ever was; I've known them both in my time. You know as well as I do that Fiat didn't need that little boutique. But Gianni listened, reasoned, and felt, with that strange Piemontese mixture of pragmatism and passion—your friend Jano had it—and decided to reach out and keep a worthy tradition alive. Otherwise, I'm sure that Ferrari would be only a memory today. So who really wrote the happy ending to your story?"

Notes and Photo Credits

The writing, editing and production of this book about Enzo Ferrari and his cars has been an ambitious project replete with the many frustrations and delights one might expect when dealing with such a difficult subject. Because of the great complexity of the man involved and the vast and confusing number of car types and engines built over nearly thirty years, opportunities for error were perhaps greater than usual.

The editor would like to extend his sincere appreciation to those whose assistance, knowledge, and involvement with the marque have done much to abet the accuracy of the book and enhance its content in many subtle but important ways.

At the Ferrari factory itself, Dr. Franco Gozzi provided rare photographs of Enzo Ferrari and Dino and Ing. Ferrari himself gave AUTOMOBILE *Quarterly* permission to reprint other photographs and data from the very latest edition of his autobiography, a work not generally available to the public at this time. Ing. Dondo should also be mentioned. His gracious cooperation with Griff Borgeson was of vital importance to several of this book's chapters.

In Italy too, the House of Pininfarina was generous with assistance and provided many photographs of historic Ferraris from its archives.

In the United States, Ferrari expert Richard F. Meritt—co-author with the late Warren Fitzgerald of *Ferrari, The Sports and Gran Turismo Cars* published by the Bond Publishing Company—was deeply involved with checking the manuscript for factual accuracy. His knowledge of the technical aspects of Ferrari cars—the engines and running gear—contributed much to the more technical sections especially.

Several Ferrari enthusiasts also deserve special mention for they gave generously of their time and shared their enthusiasm with the editor. They include: Helmut J. Brandt, Carle C. Conway, Robert N. Dusek, Paul Pappalardo, Larry Taylor and Frank Weinberg who exercised his Dino for the Mystique chapter. Bill Harrah and those who work with him were uniformly helpful as was the Ferrari importer in England, Michael Salmon.

Finally, the editor would like to thank his colleague, Beverly Rae Kimes, for her assistance, good advice and support throughout this project.

S.G.

CHAPTER 1
The Great Agitator

8-9, 31, 38-39: photographs by Julius Weitmann. 10-11, 12, 13, 16-17, 17, 18, 19: courtesy of Enzo Ferrari. 14, 15: courtesy of Griffith Borgeson. 20, 21: photographs by Corrado Millanta. 22-23, 25, 26-27, 28, 29, 32-33, 34-35, 36 left: photographs by Jesse Alexander. 37: photographs by Geoffrey Goddard. 40, 41: photographs by Neill Bruce. 16 left: photograph from the AUTOMOBILE *Quarterly* collection.

CHAPTER 2
A Certain Mystique

Photographs by Joe Bilbao.

CHAPTER 3
A Beginning—the 815

50-51, 58-59: courtesy of Peter C. Coltrin. 52, 53, 54: courtesy of Franco Zagari. Other photographs from the AUTOMOBILE *Quarterly* collection.

CHAPTER 4
The First Ferraris

86-87: courtesy of Enzo Ferrari. 88, 89: courtesy of Stan Nowak. 89: courtesy of Franco Zagari. 90-91: photograph by Dr. Vincente Alvarez. 91, 92-93: photographs by Corrado Millanta. 94-95, 101: courtesy of *The Autocar*. 98-99: from the Karl Ludvigsen collection.

CHAPTER 5
Sound & Fury—Ferrari Engines

102-103, 116 left, 118 left, 119, 125: courtesy of Griffith Borgeson. 105 above, 113 above, 118 right: courtesy of Enzo Ferrari. 111, 116-117, 117: photographs by Jesse Alexander. 105 below, 106, 107 left, 108-109, 110, 112, 113 left, 114, 115: courtesy of Richard F. Merritt. 107 right, 124, 126: from the Karl Ludvigsen collection. 113 below right, 127: photograph by Julius Weitmann.

Note: one captioning error and one typographical error in the b/w sections were discovered after the book had gone to press. The caption for the photo on page 28 below should read: Hawthorn's car for the 1958 Italian GP. The typo occurred in the caption on page 55. This should read Alberto, not Albert, Ascari.

Index

Index of Illustrations

FERRARI THE MAN, THE MACHINES WAS PRODUCED BY AUTOMOBILE QUARTERLY MAGAZINE
OTHER BOOKS PUBLISHED BY AUTOMOBILE QUARTERLY MAGAZINE:

MARQUE HISTORY BOOKS

The Buick: A Complete History by Terry B. Dunham and Lawrence R. Gustin ISBN 0-915038-19-6
Cadillac: Standard of the World by Maurice D. Hendry ISBN 0-915038-10-2
Camaro! From Challenger to Champion: The Complete History by Gary L. Witzenburg ISBN 0-915038-33-1
Corvette: America's Star-Spangled Sports Car by Karl Ludvigsen ISBN 0-915038-06-4
The Cars That Henry Ford Built by Beverly Rae Kimes ISBN 0-915038-08-0
Mustang! The Complete History of America's Pioneer Ponycar by Gary L. Witzenburg ISBN 0-915038-13-7
Opel: Wheels to the World by Karl Ludvigsen ISBN 0-915038-16-1
Packard: A History of The Motor Car and The Company edited by Beverly Rae Kimes ISBN 0-915038-12-9
Porsche: Excellence Was Expected by Karl Ludvigsen ISBN 0-915038-09-9
Scheduled: Firebird, Mercedes, American Trucks and Chevrolet

TRANSLATED EDITIONS

BMW: A History by Halwart Schrader, translated and adapted by Ron Wakefield ISBN 0-915038-15-3

RESTORATION GUIDE SERIES

The Complete Corvett Restoration & Technical Guide—Vol. 1, 1953-1962 by Noland Adams ISBN 0-915038-14-5

HOBBY BOOK SERIES

Automobile Quarterly's Complete Handbook of Automobile Hobbies edited by Beverly Rae Kimes ISBN 0-915038-28-5

COMMEMORATIVE EDITIONS

The Motorcars of Errett Lobban Cord: Auburn, Cord, Duesenberg by Griffith Borgeson ISBN 0-915038-35-8
Scheduled: Porsche, Ferrari, Mercedes, Jaguar

GENERAL

Corvette: A Piece of the Action by William L. Mitchell and Allan Girdler ISBN 0-915038-11-0
The Best Of Corvette News edited by Karl Ludvigsen ISBN 0-915038-07-2
Porsche Panorama: The First Twenty-Five Years, the official publication of The Porsche Club of America ISBN 0-915038-32-3

FERRARI POSTER

15 Beautiful Ferraris, 1947-1973
Full-color, giant size, 25 by 38 inches, $3.95 each plus $1.75 for insured shipping.

Send for brochure of over fifty other great full-color marque posters from
Automobile Quarterly Magazine and Books, 245 West Main Street, Kutztown, Pennsylvania 19530